Satellite
Communications

WILEY SERIES IN TELECOMMUNICATIONS

Donald L. Schilling, Editor
City College of New York

Satellite Communications

The First Quarter Century of Service

David W. E. Rees

INTELSAT
Washington, DC

WILEY

A Wiley-Interscience Publication

John Wiley & Sons

New York · Chichester · Brisbane · Toronto · Singapore

Library of Congress Cataloging-in-Publication Data:

Rees, David W. E.
  Satellite communications.

  (Wiley series in telecommunications)
  "A Wiley-Interscience publication."
  Bibliography: p.
  Includes index
  1. Artificial satellites in telecommunication.
I. Title. II. Series.
TK5104.R44 1989   384.5′1   88-33949
ISBN 0-471-62243-5

Printed in the United States of America

10 9 8 7 6 5 4 3 2 1

To Caryl and Ffiona

CONTENTS

The Ascension Island Apollo earth station (1966).

Man's first steps on the moon.

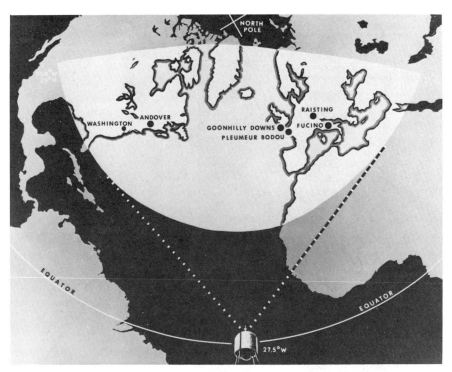

North Atlantic satellite service from INTELSAT I (Early Bird) in 1965.

The United Kingdom's Goonhilly 1 earth station (1965). Photograph provided by British Telecom International.

The United States' Andover 1 earth station in Maine.

The United States' Andover 1 earth station in Maine.

France's Pleumeur Bodou earth station complex.

Germany's Raisting earth station complex.

The Longonot earth station in Kenya (1970). One of the early Indian Ocean region users.

The ceremonial opening of the Longonot earth station in Kenya 1970.

The INTELSAT VI satellite (half-scale model).

The Titan and Ariane launch vehicles.

Small transportable earth stations.

A fly-away earth station for emergencies.

An IBS earth station in Toronto, Canada.

The Houston International Teleport.

A rural earth station in Nigeria (1977).

The Eik mainland domestic earth station in Denmark.

Domestic earth stations in Norway. The Statfjord oil rig earth station (top) and the Isfjord earth station (below).

FOREWORD

Telecommunications have undergone a revolution during the twentieth century, and especially since the first commercial use of satellite communications in 1965. The Cable & Wireless Group of companies have been a world leader in international telecommunications since the middle of the nineteenth century and remains one to this day. The story in this book concerns that part of the revolution that has resulted from satellite telecommunications and provides ideal material for people seeking an overview of how satellite communications services came to be introduced, how they expanded, and the extent to which they are used today both nationally and internationally.

I have known the author, Dai Rees, for 34 years when we first started our careers at the Porthcurno Engineering College. The intervening period turned out to be the most exciting years in telecommunications history. Dai spent many years developing telecommunications in South America, Africa, and the Caribbean before being nominated to work for INTELSAT. He became involved with satellite communications when the Cable & Wireless Group built its first earth station on Ascension Island to support NASA's Apollo Project to send a man to the moon, and he has "tracked" satellites since INTELSAT I. Before moving to INTELSAT's headquarters in Washington, D.C. he did much for the organization of early earth stations in Ascension Island, Kenya, and the Caribbean; some of the methods he devised then are still in use today. His 34 years experience in the telecommunications industry can be seen in this book, which should be read by all interested in international telecommunications.

GORDON OWEN
Joint Managing Director
Cable & Wireless Group

July 1989

FOREWORD

INTELSAT is celebrating its 25th anniversary of global telecommunications service this year. During this period of time, INTELSAT has become virtually synonymous with the phrase "live, via satellite". Yet, as millions of viewers worldwide watch televised entertainment, sporting, and breaking news events, few pause to think of the satellite system—owned and operated by a remarkable international cooperative of 117 member countries—that makes these transmissions possible. Perhaps this is so because the INTELSAT satellites' record of reliability has made exceptional service seem commonplace.

The author of this book has set out to improve public knowledge of satellite communications by describing the many INTELSAT contributions to world communications as well as those of other satellite organizations. The book provides an historical background, a summary of global satellite services, a description of providers, and a brief digest of each country's use of satellites for communication purposes.

Satellite communications have done much to benefit all countries, developed and developing, and they will continue to do so. Because we live in a rapidly changing environment, with ever-increasing telecommunications requirements, innovations have occurred even while this book was at press. For example, INTELSAT has introduced the option to lease bulk capacity for unrestricted use which will permit even greater flexibility for users to configure efficient international, regional, and domestic networks. The volume of satellite services continues to grow swiftly, especially in television distribution and basic services to smaller earth stations.

INTELSAT is extremely proud of its first 25 years of accomplishments and remains pledged to provide the high quality, reliable, and economic services for which it is known. As INTELSAT embarks on the next quarter century, its past accomplishments and future programs ensure that it will continue to play a vital role in global communications.

PEDRO J. CASTELO-BRANCO
Deputy Director-General
INTELSAT

July 1989

PREFACE

Satellite communications have, in many ways, revolutionized the way in which the world lives, works, and entertains itself. It is a wide subject and both general and very specialized books have been written on all its aspects.

What can I add to this already vast volume of books? I believe that the 23 years of my life spent on the practical aspects of satellite communications in many countries can be used to share my knowledge with those who may be already working, or starting to study and work, in this field.

This book is intended to be as nontechnical as possible and to provide a general, but essential, knowledge of how satellite communications have been, and are being applied to everyday life, as well as describing how satellite communications have spread out to the whole world.

To understand many of the things that happen today, I believe that it is necessary to know something about how we have reached the present stage, so I have described some of the recent history of how satellite communications came to be applied at all—especially the international aspects with which I have been mainly involved. Like many other things in life, satellite communications are rooted in history, which cannot be ignored. Despite what many think, all communications between people need cooperation to succeed; communication will not be successful if it is forcibly directed at a reluctant correspondent.

I hope this book will be useful to anyone involved, or likely to become involved, either as a student or in business, in the use of satellite communications, especially international communications. To make it understandable to people without a highly technical background, technical language has been kept to a minimum, although, I fear, this is not entirely possible. I have tried to describe briefly all the commercial satellite communication systems that were available in 1988 and how they are being applied. Some of the specific information will no longer be accurate by the time this book is read but this has been kept to the minimum; the intention is to describe a broad base of what is happening, not the day-to-day detail of specific applications.

Developed countries have used satellite communications to develop themselves further and to expand their economies; to do this they have used extensive, dependable, and cost-effective communications networks. To many lesser developed countries, satellites offer the only means of pursuing the same course. INTELSAT was created as an international organization to do this, as well as to improve international communications. With this in mind, about a quarter of this book describes how this has so far been achieved by individual countries. Given the continued support of its members, INTELSAT will continue this course to achieve the common goal, as will the other international organizations mentioned.

Many acknowledgments must be made; firstly to INTELSAT for permitting me to publish the book, and to reproduce many of its diagrams, photographs, and data. Secondly, I would like to thank many colleagues and friends, both inside and outside the organization for their comments and support in ensuring the accuracy of various parts of the text.

INTELSAT, and in a smaller way INMARSAT, are remarkable organizations in that they are international cooperatives, combining very fine ideals with just commercial practices. Both have provided the world with communications advantages unimaginable about half a century ago. In the last few years, INTELSAT especially, has received much adverse publicity for several (I believe unjust) reasons. As an international cooperative it has been an immense success, and this book is intended to publicize the efforts of people from many countries who have devoted their time to the advancement of world-wide cooperation through communications, whether they be part of INTELSAT, INMARSAT, or any other organization.

Lastly, but by no means least, my wife and daughter have been a constant support, not only allowing me to use their spare time, but also in putting up with the constant typing and other inconveniences that go with writing a book!

DISCLAIMER

The views, opinions, and judgements expressed in this book are solely those of the author, and do not necessarily reflect the official policies of INTELSAT, its members, users, or Signatories.

DAVID W. E. REES

Vienna, Virginia,
May 1989

ACKNOWLEDGMENTS

The Management of INTELSAT for permitting me to publish the book
William Wu, INTELSAT, for persuading me to start the book!
Regional C. Westlake, past Deputy Director General of INTELSAT
Frank L. Miklozek, past President of the National Association of Post-
masters of the United States
Baroness Lucy Faithfull, House of Lords, London
Michael R. Sweeney, Senior Lecturer, Cranfield Management School,
England.
All are greatly appreciated for their encouragement and advice.

D. W. E. R.

Satellite
Communications

1

SATELLITE COMMUNICATIONS— THE NEED

The need for human beings to communicate and the means for communicating have evolved over time. In the more recent realms of human advancement, the continuing discovery of mankind's terrestrial and spatial environment has created a need for a wider variety of better and faster communications over long distances—telecommunications.

For about four centuries after the dramatic age of discovery when Columbus determined that the world was round, and when Vasco Da Gama, Hudson, Drake, and others discovered so much more of the world's surface, the only means for seafarers to communicate back to their home countries was to leave a letter in a post-box tree, or to send an overland message if that were possible. It was not until the industrial revolution in the 19th century that electric means of communicating became possible. These inventions became known as the telegraph (far-writing by electricity) and the telephone (far-speaking by electricity) for written and spoken communications over a distance, both known as telecommunications.

The advent of space communications using satellites in the early 1960s and the formation of an international cooperative organization (INTELSAT) to develop and provide global telecommunications has completely revolutionized the world's ability to communicate. Founded with high ideals, which have been sustained during its first quarter century, INTELSAT has stood the test of time. It has developed into one of the technological wonders of the world promoting the peaceful use of outer space for telecommunications in the most efficient way possible, and making available these resources on a shared and equitable basis for all of mankind at the least possible cost.

However, before describing the development and present status of global satellite communications services, it is beneficial to consider some of the earlier developments in the field of telecommunications.

1

The letter box tree at Mossel Bay in South Africa near the Cape of Good Hope.

1.1.   EARLIER DEVELOPMENTS IN TELECOMMUNICATIONS

Through technological advancement, the distance over which humans can communicate has become greater, and it is a worthwhile exercise to refresh our memory on how recently some of these events have taken place. Some of the more important events are as follows:

1837 – A patent was granted, in England, to Charles Wheatstone for an electric telegraph system. The Cooke and Wheatstone Electric Telegraph Company was formed in the United Kingdom which worked over 1.5 km and became the world's first electric telegraph system. The company later constructed a 21 km telegraph line in West London.

1844 – Samuel Morse demonstrated his telegraph code, the Morse code, over a telegraph line in the United States between Washington D.C. and Baltimore.

1845 – The General Oceanic Telegraph Company was registered in New York to link Europe and North America.

1849 – The first telegraph cable was laid between England and France although it broke down after only 8 days.

1851 – The International Electric Telegraph Commission met in Paris, France and it is interesting to note that of the four countries forming this commission (United Kingdom, France, Prussia and Belgium), United Kingdom was the only country in which the new services were not operated by a government department of Posts and Telegraphs.

1851 – The first commercial submarine cable service in the world was started between England and France, by private enterprise.

The mid-1800s was a period of private enterprise, especially in the United States and the United Kingdom. In the United Kingdom, the government had decided that the Post Office would take over the country's internal telegraphs as soon as it was sure of their reliability, but presumably horrified by the high investment cost and the risks involved in the new technology, left the building and operation of United Kingdom's external services to publicly floated companies.

A joint Anglo-American company was formed by two entrepreneurs (Cyrus Field of New York, and John Pender of London) called The New York, Newfoundland, and London Telegraph Company to link North America with the United Kingdom. Cyrus Field had problems in raising funds for his venture in New York but had better success in London so he transferred his charter to an all-British company named The Atlantic Telegraph Company. Further funds became available from the British and American governments, and after some inherent setbacks, the first transatlantic messages were exchanged on 18 August 1858. The cable was short-lived, and the signals deteriorated quickly, but the future benefits for such communications were clearly seen.

In retrospect, it is interesting to observe the similar events that took place in telecommunications during the 1850s and about 100 years later when satellites were introduced for the first time in the mid-1960s.

The next 100 years between 1855 and 1965 saw many technological advances in communications and a consequent explosion in services extending from the North Atlantic to all parts of the world. Again, some of the outstanding feats and dates are shown so that the reader can understand how satellite telecommunications services came to be needed 100 years later.

1861 – The United States transcontinental telegraph cable was completed.

1868 – The first commercial transatlantic telegraph cable was completed between the United Kingdom and Canada with onward land connections to the United States.

1876 – In the United States a telephone patent was issued to Alexander Graham Bell who first transmitted speech on 10th March.

1901 – Marconi transmitted the first transatlantic wireless message from the United Kingdom to Canada.

1902 – Round-the-world telegraph service was started using cables laid by the forerunners of the Cable & Wireless Group.

1926 – R. H. Goddard successfully launched a rocket in the United States and J. Baird first demonstrated television in the United Kingdom.

1927 – The first commercial transatlantic radio telephone service was started.

1929 – The British Broadcasting Corporation (BBC) started experimental TV transmissions from its London studios. In the United States, a new private company was formed to operate aeronautical services.

1936 – The first regular television services started in the United Kingdom and Germany. Experimental color transmissions were made in the United Kingdom.

These advancements allowed entrepreneurs of the time to create some highly successful long distance telecommunications companies, many of whose names have long since been forgotten. Many of these companies dissolved or amalgamated over the years but it is noteworthy that many of them were privately owned having such names as

The Eastern Telegraph Company
The West Coast of America Telegraph Company
The Marconi Wireless Telegraph Company
Western Union
France Cables et Radio
Cable & Wireless
Italcable
All America Cable & Radio

Competition became hard and fast between the companies and between the technologies causing many to be taken over, some to amalgamate, and in some cases, become large successful companies. Some of the best known survivors are companies whose names are still well known today such as The American Telephone and Telegraph (AT&T), Western Union, Cable & Wireless, France Cables, Italcable, Radio Corporation of America (RCA), and others.

These companies consolidated but there was no radical technological change until the 1950s when coaxial telephone cables first appeared. Now, in 1988, it is interesting to recall the telecommunications services that were available to a typical business only 30 years ago in 1958. Most developed countries had national telegraph, telephone, and possibly facsimile services; some had subscriber telex, but the usual means of sending a message was to take it to a telegraph company (private or nationalized depending on the country) which would pass it through its network and then send it by messenger to the addressee.

At that time I was working in a South American capital city with one of the leading telecommunications companies of the time using some equipment which dated to the turn of the century. There were peak traffic times just as we have today; one of the peak periods was around lunchtime when businessmen handed in their urgent cables on their way to lunch knowing that the reply from New York would be delivered to their desk when they returned from lunch two or three hours later.

International telephone calls were either difficult or impossible in 1955. All intercontinental telephone calls were forced to use the high frequency radio bands which were congested and noisy. During the late 1950s, the world's leading nations were starting to explore beyond the earth's terrestrial surface which, to be successful, required a great step forward in mankind's ability to communicate.

The concept of "extraterrestrial relays" had first been advanced by Arthur C. Clarke in the British magazine *Wireless World* in October 1945. This article theorized that if three radio relay stations were positioned in space at equidistant locations above the equator, then global communications would be possible. An extract from that article, referring to problems of communications in 1945, is given below.

All these problems can be solved by the use of a chain of space-stations with an orbital period of 24 hours, which would require them to be at a distance of 42,000 km from the centre of the earth [Figure 1.1]. There are a number of possible arrangements for such a chain [Figure 1.2], but that shown is the simplest. The stations would lie in the earth's equatorial plane and would thus always remain fixed in the same spots in the sky, from the point of view of terrestrial observers. Unlike all other heavenly bodies they would never rise or set. This would greatly simplify the use of directive receivers installed on earth.

The following longitudes are provisionally suggested for the stations to provide the best service to the inhabited portions of the globe, though all parts of the planet will be covered.

    30°E   Africa and Europe
    150°E   China and Oceania
    90°W   The Americas

Each station would broadcast programmes over about a third of the planet. Assuming the use of a frequency of 3000 megacycles, a reflector only a few feet across would give a beam so directive that almost all the power would be concentrated on the earth. Arrays a metre or so in diameter could be used to illuminate single countries if a more restrictive service was required.

The stations would be connected with each other by very-narrow-beam, low-power links, probably working in the optical spectrum or near it, so that beams less than a degree wide could be produced.

The system would provide the following services which cannot be realized in any other manner:

    a. Simultaneous television broadcast to the entire globe, including services to aircraft
    b. Relaying of programs between distant parts of the planet

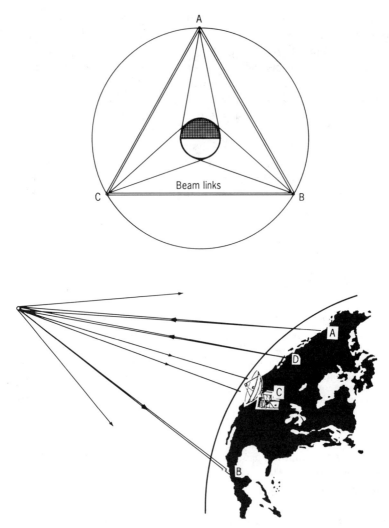

Figure 1.1. Arthur Clarke's geosynchronous orbit.

Of course, at that time there were no means available for launching such a relay station, as the largest rockets in 1945 were only capable of launching bombs from continental Europe to the United Kingdom. It was to take 10 more years before the next major breakthrough took place in the advancement of telecommunications, and another 17 years before Arthur Clarke's theory became a reality. In the intervening years:

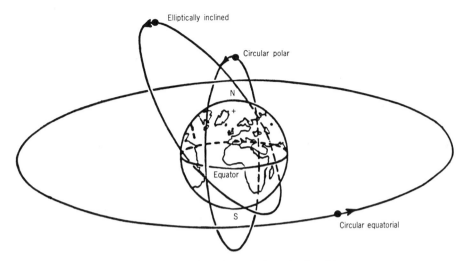

Figure 1.2. Arthur Clarke's three basic orbits.

1956 – TAT1 (Trans-Atlantic Telephone cable Number 1) started service between the United Kingdom and Canada with 36 circuits and using a bandwidth of 552 MHz.

1957 – The Union of Soviet Socialist Republics (U.S.S.R.) launched the first SPUTNIK satellite.

1958 – The United States launched its Explorer 1 satellite.

1959 – TAT2 cable augmented service between Canada and France while Explorer 2 provided the first TV picture from space via satellite.

1960 – The United States launched an Echo satellite providing the first self-powered communications satellite, and another satellite, Courier 1 was the first satellite with solar cells.

1962 – CANTAT-2 provided 80 new telephone circuits between Canada and the United Kingdom and a new satellite, TELSTAR-1 was launched providing the first commercially funded communications satellite over the Atlantic ocean. President John F. Kennedy signed the Communications Satellite Act in the United States. In December 1962, the European Conference of Posts & Telecommunications (CEPT) was formed to strengthen West European telecommunications policy.

The years 1962, 1963, and 1964 were hectic years for transatlantic telecommunications as two more telephone cables were laid, and a number of communications satellites were launched; 1964 also saw the first transpacific cable with 138 circuits and the birth of the International Telecommunications Satellite Consortium known as INTELSAT.

TABLE 1.1 Major Satellite Series Launched

International Satellites

Year	Satellite	Number in the Series
1965	INTELSAT I (Early Bird)	1
1967	INTELSAT II	4
1968	INTELSAT III	8
1971	INTELSAT IV	7
1975	INTELSAT IV-A	6
1976	MARISAT (for INMARSAT)	3
1981	MARECS (for INMARSAT)	2
	INTELSAT V	9
1984	EUTELSAT I	5
1985	ARABSAT	2
	INTELSAT V-A	6
1988	PanAmSat (private U.S. company)	1
	INMARSAT II	

National Satellites

Year	Country	Satellite
1957	U.S.S.R.	Sputnik I
1958	United States	Explorer 1
1960	United States	Echo
1962	United States	TELSTAR (AT&T)
1962	United States	RELAY (RCA)
1963	United States	SYNCOM (Hughes Aircraft)
1972	Canada	ANIK A
1974	United States	WESTAR (Western Union)
1975	United States	SATCOM (RCA)
	U.S.S.R.	Raduga
1976	United States	COMSTAR (COMSAT General)
	Indonesia	Palapa
	U.S.S.R.	Elkran
1978	U.S.S.R.	Gorizont
1980	United States	SBS (Satellite Business Systems)
1981	United States	Aurora (Alascom)
	United States	GStar (GTE)
1983	Japan	CS-2
	United States	Galaxy (Hughes Communications)
	United States	Telstar 3 (AT&T)
	United States	Leasat (Hughes Communications)
	United States	TDRSS (NASA, experimental)
	India	Insat
1984	United States	Spacenet (GTE)
	France	Telecom
	Japan	BS-1
1985	Australia	AUSSAT
	Brazil	Brazilsat
	Mexico	Morelos
	United States	AMSAT (Continental Telephone)
	United States	Satcom-K1
1987	Germany	TV-SAT
1988	France	TDF
	Luxembourg	Astra

TABLE 1.1   (*Continued*)

	Future launches scheduled	
	International Satellites	
Year	Satellite	
1989	INTELSAT VI	
	EUTELSAT II	
	INMARSAT II	
1990	INTERSPUTNIK-TOR	
1992	INTELSAT VII	
1994	INMARSAT III	

	National Satellites	
Year	Country	Satellite
1989	United States	DBS-SAT
	Italy	Italsat
	United Kingdom	BSB
	Canada	ANIK E
	Sweden	TELE-X
	United Kingdom	ASIASAT
1991	Tonga	
	Iran	

The U.S. Apollo Program was the first commercial application of satellite communications. In effect, this was the first round-the-world business telecommunications network using satellites and interconnected earth stations in different countries; the networks being introduced today by many large corporations and international organizations are remarkably similar.

The National Aeronautics and Space Administration (NASA) required a worldwide telephone and data communications network to link its headquarters and launch facilities in the United States with its orbiting satellites and the moon via earth stations located strategically in foreign countries. This network included the Goddard Space Flight Control Center in Maryland, Communications Satellite Corporation (COMSAT) earth stations in Maine, California, and Hawaii, NASA tracking stations overseas operated by foreign entities, and also tracking ships on the high seas.

It was at one of these earth stations on the United Kingdom's Ascension Island, strategically situated in the middle of the South Atlantic, that I first became involved in satellite communications. NASA had installed a large tracking installation on the island, and the commercial telecommunications operator, Cable & Wireless, installed its first commercial earth station there in 1966 to provide the necessary voice and data communications between the tracking station and the United States. The NASA Apollo network was the first international satellite business data network and made the U.S. Apollo mission to place Man on the Moon possible. It was a very exciting time in the history of telecommunications.

Since that time, many telecommunications satellites have been launched both for national and international services which have permitted an unprecedented growth in the communications between humans and between the machines used by humans. The remainder of this book describes many of the enormous range of applications now available for the human race by means of an ever-increasing number of satellites located in the geosynchronous orbit 23,000 miles over the earth's equator. Chronological lists of some of the major international and national satellite series of launched are shown in Table 1.1. A more comprehensive list of commercial satellite launches is provided in Appendix C.

1.2. THE FORMATION OF INTELSAT

The need for an organization such as INTELSAT arose from the U.S. space program, in particular the NASA Apollo Program aimed at putting man on the moon within the decade, with regular manned space flights. These programs could only be completed successfully if there were reliable first class telecommunications services between the spacecraft and the ground controls in the United States. Such services did not exist and could not be implemented in any way other than using telecommunications satellites and earth stations in a number of strategic places round the world. There was thus the technical need for satellites and a political need to provide the service internationally very quickly, efficiently, and cooperatively. It may also be added that there were many business interests at stake supporting the space program and all its expected benefits to the United States.

International negotiations actually started in 1959 when the World Administrative Radio Conference (WARC/1959) of the International Telecommunications Union (ITU) took preliminary steps by allocating some frequency bands for space communications research. This was followed by bilateral agreements between the American, French, British, and later other governments to build earth stations.

The 1960s and 1970s were exciting times in space exploration and telecommunications. Many conferences were held and many articles written on satellite communications issues. One of the most independent and comprehensive references of the time was Judith Tegger Kildow's *INTELSAT: Policy Maker's Dilemma* in 1973. It resulted from a study performed for the Massachusetts Institute of Technology, and I will be quoting several passages from it.

The United States Congress passed the Communications Satellite Act in August 1962 with the prime objective of instituting an international telecommunications network as soon as possible. COMSAT was incorporated as a private corporation in February 1963, with a mixture of public and private objectives.

The act stressed commercial objectives but permitted flexibility for political considerations, especially with regard to international participation.

To many, COMSAT was born with a conflict of interests; the act states, among other things, that:

Section 102(a)

The Congress hereby declares that it is the policy of the United States to establish, in conjunction and in cooperation with other countries, as expeditiously as practicable, a commercial communications satellite system, as part of an improved global communications network, which will be responsible to public needs and national objectives, which will serve the communications needs of the United States and other countries, and which will contribute to world peace and understanding.

Section 102(b)

The new and expanded telecommunication services are to be made available as promptly as possible, and are to be extended to provide global coverage at the earliest practicable date. In effectuating this program, care and attention will be directed toward providing such services to economically less developed countries and areas as well as those more highly developed, toward efficient and economical use of the electromagnetic frequency spectrum, and toward the reflection of the benefits of this new technology in both quality of services and charges for such services.

Section 102(c)

In order to facilitate this development and to provide for the widest possible participation by private enterprise, United States participation in the global system shall be in the form of a private corporation, subject to appropriate governmental regulation.

"The dual philosophy of the Act was the basis for conflict and confusion between public and private interests, national and international interests" (Kildow 1973: 42, 43). The United States, and the rest of the world, badly needed a reliable international telecommunications network for which only satellites could be used. The State Department had international political interests; other government agencies had their specific interests, and lastly, but by no means least, private business interests were abundant.

COMSAT's powers were laid out in the act so that it could operate with authority, but the powers given to other U.S. departments and agencies were much less defined and later led to much ambiguity. Some of the powers given to COMSAT are provided in Section 305 of the act including:

(a) In order to achieve the objectives and to carry out the purposes and powers of this Act, the corporation is authorized to
   1. Plan, initiate, construct, own, manage, and operate itself, or in

conjunction with foreign governments or business entities, a commercial communications satellite system;

2. Furnish, for hire, channels of communications to the United States common carriers and to other authorized entities, foreign and domestic; and

3. Own, and operate satellite terminal stations when licensed by the Commission (FCC) . . .

(b) Included in the activities authorized to the corporation for accomplishment of the purposes indicated in subsection (a) of this section are, among others not specifically named:

1. To conduct or contract for research and development related to its mission;

2. To acquire the physical facilities, equipment and devices necessary to its operation, including communications satellites and associated equipment and facilities, whether by construction, purchase, or gift;

3. To purchase satellite launching and related services from U.S. Government;

4. To contract with authorized users, including the United States Government, for the services of the communications satellite system; and

5. To develop plans for the technical specifications of all elements of the communications satellite system.

(c) To carry out the foregoing purposes, the corporation shall have the usual powers conferred upon a stock corporation by the District of Columbia Business Corporation Act.

COMSAT got its powers from the U.S. government to determine policies from the State Department, and to determine rates from the Federal Communications Commission (FCC). It was also bound by U.S. corporate law. This was an extraordinary situation with COMSAT given broad powers to organize an international telecommunications system, but at the same time having to confer with the State Department, the FCC, and also be scrutinized by other presidential and government agencies.

In accordance with the United Nations General Assembly Resolution 1721 on the peaceful uses of outer space, worldwide telecommunications representatives from 19 countries (Australia, Austria, Belgium, Canada, Denmark, France, Germany, Ireland, Italy, Japan, the Netherlands, Norway, Portugal, Spain, Sweden, Switzerland, United Kingdom, United States, and the Vatican) met in Washington during July 1964 to participate in the International Plenipotentiary Conference on Interim Arrangements for a Global Commercial Satellite System. The three principal issues were the organizational structure, its management, and whether other systems would be permissible.

Although the U.S.S.R. had participated in some earlier negotiations, they did not participate in the INTELSAT negotiations. Although reasons were never announced, this was probably due to the fact that the U.S.S.R. wished to exploit space for noncommercial reasons, whereas the United

States wanted to exploit space for commercial purposes. "European governments, accustomed to government owned and operated communications entities, were reluctant to allow technicians and businessmen to negotiate such important foreign policy matters and insisted at the very least that public officials participate" (Kildow 1973: 44, 45).

The United States' primary objective was to establish a satellite system as rapidly as possible. The prime objective of other countries was to create an international organization as soon as possible, with a management structure or secretariat. The Europeans wished to participate in the organization's management whereas the United States' representative (COMSAT) wanted to manage the system.

All agreed on the principle of a single global commercial system, but views differed as to whether other systems should be prevented. The interim solution was that:

a. An Interim Communications Satellite Committee (ICSC) would be formed to be the temporary governing body.
b. The United States' COMSAT would be appointed manager of the system during the period of the Interim Agreements.

The conference concluded on 20 August 1964, when 11 countries agreed the adoption of two Interim Agreements for a global commercial satellite system:

1. An agreement among governments establishing interim arrangements for a global commercial communications system
2. A special agreement, signed by the designated communications entities of each nation, setting up the mechanics to carry out the interim arrangements

In effect, INTELSAT was born on that date—20 August 1964—the day on which those agreements were opened for signature and were signed by the first seven countries (Canada, Japan, Netherlands, Spain, United Kingdom, United States, and the Vatican).

The United States Government's intentions for the new organization were stated by President Johnson in the inaugural television transmission using the Early Bird satellite on 28 June 1965.

This moment marks a milestone in the history of communications between peoples and nations. For the first time, a manmade satellite of earth is being put into commercial service as a means of communication between continents. The occasion is as happy as it is historic, and that is for many reasons.

This is first of all, a very tangible and valuable realization of the promise and potential of man's exploration of space. On ahead, we shall take many more and, I think, many longer strides forward. But we can know, from this step today, that mankind's growing knowledge of space will bring growing improvement for life on earth. So, it is especially fitting that this historic step comes not as the achievement of any single nation—but as the work of many nations. This represents a joint venture of 44 countries, with still more participants in prospect. For us in the United States, that is especially gratifying. Since the earliest days of the age of space we have urged—as we still do—that all nations join together to explore space together and to develop together its peaceful uses.

Finally, for us—and, I am sure, for our friends in Europe as well—it is a particularly happy circumstance that this service is another bond in the many ties that join us together across the North Atlantic.

Other satellites, in the days to come, will open new communications pathways for all the world, but we are especially pleased that this first service brings closer together lands and people who share not only a common heritage, but a common destiny—and a common determination to preserve peace, to uphold freedom, to achieve together a just and a decent society for all mankind.

In these times, the choice of mankind is a very clear choice between cooperation and catastrophe. Cooperation begins in the better understanding that better communications bring. On this occasion, then, I am pleased to extend my congratulations to all the international participants in this system and to the Communications Satellite Corporation, and I would express the hope that all nations may become willing to join in such great enterprises for the good of mankind, and that all our labors may be blessed by a rich and a bountiful harvest of peace on earth.

1.2.1. The Formative Years 1964–1973

Once the INTELSAT system was established, at least on an interim basis, more countries joined the new organization and a review was made of the system's effectiveness before the final definitive arrangements were considered. Not surprisingly, much of the action and policymaking during this period took place within the United States and centered around the U.S. representative, and INTELSAT's manager, COMSAT.

With the management of the system, COMSAT controlled the economic and legal components of its operations. As manager, COMSAT was authorized to place contracts relating to the design and development of the system, and was in charge of all equipment procurement, only requiring ICSC agreement on major decisions which was not difficult to obtain being the U.S. representative and having over 60% of the investment share. With this decision making power, COMSAT had the power of initiatve in recommending courses of action for the entire system as well as day-to-day operations. "The controversy over U.S. dominance of the system, basing ownership and voting power on expected usage, again reflected the conflicting values between the underlying commercial philosophy espoused by the U.S. and that of equitable political participation advocated by others" (Kildow 1973: 49). During this period COMSAT came under increasing worldwide criticism for its monopolistic tendencies within the United States in being the sole satellite service provider, and for its coercive approach towards other countries abroad. On account of these concerns, President Johnson in 1968, urged a continuation of the Interim Agreements and established a task force to study the overall United States policy on telecommunications—the Office of Telecommunications Policy (OTP). It was President Nixon who staffed the OTP in 1969 that was responsible for the final policy for the INTELSAT Agreements on behalf of the United States.

In the United States the existing telecommunications companies, broadcasting, television, educational, and other interests were not all happy with the creation of a new corporation, mainly headed by scientists, lawyers, and retired military, to become responsible for the whole new field opening up for satellite communications. AT&T, together with other U.S. carriers applied to the FCC to lay another transatlantic cable (TAT-5) and COMSAT objected to this as unnecessary competition to the new satellite business.

Overseas, COMSAT as manager for the network, controlled the operational aspects of the embryo INTELSAT and although they earned a good deal of respect for their organizing ability, their monopolistic approach was not really popular with other members. "The Europeans rejected the idea of COMSAT as Manager (on a permanent basis), suggesting that Europe must participate in management. They proposed an 'in-house' staff to carry out management functions. Later, while the Interim Arrangements were still in effect, they proposed that if COMSAT were Manager, then non-Americans be assigned to the COMSAT staff." (Kildow 1973: 54) COMSAT executives feared "that an international management would take INTELSAT down the UN path—jobs filled according to national origin." (Kildow 1973: 54)

Some, but not all, U.S. policy makers saw INTELSAT as an American controlled organization, and the unyielding U.S. position during the Interim Negotiations for the Definitive Arrangements (1965–1973) showed that "Congress intended full control by COMSAT over such matters and COMSAT did not envisage a multilateral international organization. Hence, the underlying commercial philosophy of COMSAT Corporation was destined to be transferred to INTELSAT. Those who were in power would remain in power." (Kildow 1973: 10)

The original presidential policy of a cooperative organization through legal, economic, and political means, was changed to that of a U.S.-dominated organization which in turn encouraged other countries to further their own interests through the organization. A unique opportunity for international cooperation was already being diminished. Neither Congress nor COMSAT grasped the true international significance of INTELSAT. "While President Kennedy spoke of the 'U.S. portion of the system,' most Congressmen saw the proposed global system as a 'U.S. show,' and the COMSAT officers saw it as COMSAT's show. The situation has been compared to that of the Union Pacific Railroad 'Anyone could ride, but it was our railroad'." (Kildow 1973: 9, quoting Abram Chayes' *Unilateralism in U.S. Satellite Communications Policy*). Quoting from the conclusions of Judith Kildow's (1973) book:

> Perhaps the United States could have worked out better arrangements for its ambassador to INTELSAT. Caught up in the momentum of the need for technical expertise and the pressures of the business community, the United States Government—whether inadvertently or not—entrusted its foreign policy-making duties to a private corporation. And what were the results of COMSAT's decisions?
>
> There is one school of thought that believes that international organizations should be neat, understandable packages, in proper legal order. This perspective would have liked to have seen INTELSAT as a comprehensive legal entity, coordinating and operating all functions of communications satellites, including air and sea navigation. Because INTELSAT cannot enter new fields without the consent of the Consortium, bold planning for a comprehensive and integrated system will not be possible, according to one author INTELSAT will settle into the accepted and conventional role of an intercontinental carrier of commercial voice and message traffic.... but future developments of satellite communications will occur outside the framework of the consortium.
>
> Others see international organizations as specific functional units performing particular tasks as required by the international community. While this may result in a number of not necessarily well-coordinated functional units, and a not very well-organized legal order, it is believed that the need to coordinate becomes imperative as each unit depends on something another unit has to offer and that such a functional arrangement often can work more efficiently than an artificial legal structure—which in theory should operate smoothly, but is not necessarily suited to the task. Functionally specific units and comprehensive legal entities are often incompatible. When both concepts are expected from one organization, conflicts occur.

Neither view seems to imagine anything that was not entirely American; the first seems to envisage purely an international satellite carrier (which it probably is, but it is mainly owned by governments), and the second seems to envisage an American government-controlled bureaucracy.

An early example of this determination to dominate the new organization came with the purchase of the INTELSAT III spacecraft.

> In February 1966, COMSAT asked the FCC for permission to give TRW Inc. a contract for building the third generation satellites, the 1200 circuit INTELSAT III. Yet, without waiting for a reply from the notoriously slow Federal Communications Commission (FCC), COMSAT in April 1966 went ahead and asked the members of the ICSC to approve the contract. But there was a snag. The Hughes Aircraft Company, which had made Early Bird and the second

generation of satellites (INTELSAT II), was angry at being deprived of the next big contract and complained to the FCC. The FCC then reminded COMSAT that it had not given it permission to move ahead with the contract. COMSAT was then in the awkward position of having to tell the ICSC that the contract would have to await the FCC's approval. The international partners were furious; what kind of international organization was it that owned the world's satellite system, yet could not act without the consent of a domestic agency of the U.S. Government.

—(*New Scientist*, 1 June 1972, Brenda Maddox, INTELSAT, Lament for a Lost Hope)

The United States possessed the technology that had been developed with many millions of dollars from public funds. Only the United States and U.S.S.R. could launch satellites and the United States was far ahead in the field of communications satellites. More than 80 percent of the traffic originated in the United States, and the only thing that Europe represented was the other end of the circuit. The United States was obviously very keen to exploit its strength while the situation was this favorable. The United States was generous to share the global system with others, but some of its representatives were reluctant and not prepared to share the benefits to be gained from further industrial developments. By sharing only the other end of the circuits, the United States hoped that the other countries would be content with continued American control; with the control would remain the power and profit from industrial benefits.

The Europeans did not like this arrangement, and their viewpoint was clearly expressed in 1968 at a U.S.–European space conference in Germany by Ludwig Boelkow:

> Since agreement on a European basis could not be reached about the development of a communications satellite, France and Germany decided to go ahead with their bilateral programme, Symphonie. The United States sees in this project a rival to INTELSAT. From statements made in American official circles, one can infer that the U.S. would like to see global tele-communications by satellite under the control of INTELSAT, in which they have the majority and which for all intents and purposes is entirely under American influence.

It was not therefore surprising that when negotiations were resumed in February 1969 by the new Nixon administration, they broke off only weeks later due to the inability of its members to agree on basic issues— particularly on the extent of United States control (Kildow 1973: 4). The United States needed and wanted to continue the Interim Arrangements, and was able to do so only by continually refusing foreign demands and by offering unsatisfactory compromises to the other member nations. The Definitive Agreements were opened for general signature on 20 August 1971, seven years after the entry into force of the Interim Agreements.

1.2.2. The INTELSAT Definitive Arrangements

Fortunately many of the problems foreseen during the Interim Negotiations did not materialize, and after 17 years in use the Definitive Arrangements have proved to be fine guidelines for an exceptionally successful international venture. They are based upon two agreements:

> The Agreement to which a country's government must accede when becoming a party to the Agreement.
>
> The Operating Agreement that must be signed either by a country's government, or its designated telecommunications entity which then becomes the Signatory to the Operating Agreement.

Upon deposit of an instrument of accession to the Agreement, a government becomes a party to it, but membership in INTELSAT is not achieved until the Operating Agreement has been signed by the entity designated to do so by the new party. The definitive arrangements are contained in two documents known as The Agreement and The Operating Agreement which are kept in the custody of the U.S. Department of State.

The preamble to the Agreement notes that a global commercial telecommunications satellite system exists and expresses three fundamental principles underlying the formation of INTELSAT:

- A desire to continue the development of the telecommunications satellite system with the aim of achieving a single global commercial telecommunications satellite system as part of an improved global telecommunications network which will provide expanded telecommunications services to all areas of the world and which will contribute to world peace and understanding.
- A determination to provide, for the benefit of all mankind, through the most advanced technology available, the most efficient and economic facilities possible consistent with the best and most equitable use of the radio frequency spectrum and of orbital space.
- A belief that satellite telecommunications should be organized in such a way as to permit all peoples to have access to the global satellite system and to invest in the system with consequent participation in the design, development, construction, including the provision of equipment, establishment, operation, maintenance and ownership of the system.

In Article II, the main purpose of INTELSAT is defined as "to continue and carry forward on a definitive basis the design, development, construction, establishment, operation and maintenance of the space segment of the global commercial telecommunications satellite system." Further amplification is given in Article III where the scope of INTELSAT activities is defined. This is a particularly important and interesting article. INTEL-

SAT's prime objective is described as "the provision, on a commercial basis, of the space segment required for international public telecommunications services of high quality and reliability to be available on a non-discriminatory basis to all areas of the world." The article goes on to state that the INTELSAT space segment may also be used to provide domestic telecommunications on the same basis as international services, when countries' boundaries include such obstacles as separated parts, seas, or other major barriers. This means that the use of those domestic services can increase a signatory's investment share for the purposes of the composition and the voting participation on the Board of Governors.

Other domestic telecommunications services, outside of these categories, can also be provided by INTELSAT, but in such cases the agreement requires that the provision of these services does not impair the ability of INTELSAT to achieve its prime objective. Specialized telecommunications, international or domestic, other than for military purposes, may also be provided subject to certain conditions.

INTELSAT may also provide satellites or associated facilities separate from the INTELSAT space segment, subject to appropriate terms and conditions, provided the efficient and economic operation of the INTELSAT space segment is not unfavorably affected in any way.

The financial principles of INTELSAT are defined in Article V of the agreement and are detailed in the Operating Agreement. Each member holds an investment share based upon its use of the system, subject to a minimum of 0.05 percent. (At 1 March 1988 the valuation was about US$1.5 billion and 0.05 percent about US$740,000. The revenues of the organization come from utilization charges, and after deduction of operating costs are distributed to members in proportion to their investment share as repayment of their investment and as compensation for the use of capital.

The structure of INTELSAT consists of an Assembly of Parties of the Agreement, a Meeting of Signatories to the Operating Agreement, a Board of Governors, and an Executive Organ headed by a Director General, with headquarters in Washington D.C. The structure of INTELSAT is shown in Figure 1.3 (see p. 28) and defined in Articles VI to XI of the agreement.

Any country that is a member of the ITU may become a member of INTELSAT.

The Assembly of Parties is composed of all governments that are parties to the agreement. It considers those aspects of interest to the governments, as well as the resolutions, recommendations or views addressed to it by either the Meeting of Signatories or the Board of Governors related to the general policies and long-term objectives of INTELSAT, and each party has one vote at the meetings, which normally take place every two years.

The Meeting of Signatories is composed of all Signatories to the Operating Agreement—governments or their designated telecommunications entities. It considers resolutions, recommendations or views addressed to it either by the Assembly of Parties or the Board of Governors, and also considers matters relating to the financial, technical, and operational aspects. Normally, there is one Meeting of Signatories each year.

The Board of Governors is very similar to a company's board of directors. The board is composed of those signatories whose investment shares, individually or in groups, are not less than a specified amount which is determined annually by the Meeting of Signatories. There are approximately 20 governors excluding groups combining their investment share, and groups of members who are eligible on a regional basis (of which there may be five) representing a group of at least five signatories all within the same ITU region. The board is responsible for all decisions concerning the design, development, construction, operation and maintenance of the space segment as well as for implementing any determinations that may be made by the Assembly of Parties or the Meeting of Signatories. The board is assisted by Advisory Committees, the Director General and the Executive Organ. The board endeavors to take decisions unanimously, but failing this, does so by a weighted vote based on investment shares.

The Executive Organ, initially headed by a Secretary General, is now headed by a Director General and based in Washington D.C. in the United States and is responsible to the Board of Governors for the day-to-day management of INTELSAT. The first Director General was Mr. Santiago Astrain of Chile from 30 December 1976 until 30 December 1983. Mr. Astrain was responsible for INTELSAT during the transitional period when COMSAT changed from the role of Manager to the role of a Management Services Contractor (MSC) until the Executive Organ became fully independent on 1 January 1979. During this period INTELSAT's growth was impressive and the organization became truly international in nature.

INTELSAT's second Director General was appointed in 1983 following a search and selection process defined in Article XI of the agreement. Candidates were nominated by the United States, Canada, Thailand, Algeria, and Australia.

Strong support was provided by the U.S. signatory (COMSAT) and the U.S. candidate, Richard R. Colino, was selected by the Board of Governors which was later confirmed by the Assembly of Parties in October 1983. Richard Colino took up his position on 31 December 1983. He had previously been with COMSAT between 1964 and 1980 and had been largely responsible for COMSAT's policy towards INTELSAT during the formative years, and had been Chairman of INTELSAT's Board of Governors during the period of handover of COMSAT's management control of INTELSAT to the new Executive Organ in 1976–1977.

Due to actions taken by staff within the organization, external auditors, and outside legal counsel, on 24 November 1986, the Chairman of the Board of Governors announced to the press that the Director General and his second in command, José Alegrett, had been placed on "administrative leave." Within two weeks the Board of Governors dismissed both top executives of the Organization for a number of serious financial irregularities and at the same time asked assistance from the U.S. Justice Department in continuing the investigation. Richard Colino later pleaded guilty and was sentenced to six years imprisonment, and his deputy, José Alegrett, returned to his native Venezuela until arrested in Aruba and returned to the United Sates in July 1988.

Following a further election in March 1987, Richard Colino was succeeded by a new U.S. nominee, Dean Burch, chosen from the nominees of Australia, Brazil, and Finland. INTELSAT and 110 of its 111 member States owe much to the United States for the development of its worldwide system.

Some of the foregoing concerning the United States' desire to control the organization must be offset by the fact that its signatory to the organization (COMSAT) did create a brilliantly successful system, and the U.S. Government did authorize NASA to provide launch services at basic cost with nothing added for the enormous development cost which was borne entirely by the U.S. taxpayer; without these major contributions, worldwide communications would not be where they are today. The problem of COMSAT's monopoly of international satellite services within the United States has diminished as a result of the 1984 deregulatory policies. There are now dozens of companies providing earth stations and international satellite access services in the United Sates with a noticeable reduction in end-user costs. The deregulatory policies of the U.S. Government are now beginning to take effect, be reflected in the actions of some other governments, and are having some beneficial effects on the expansion of international telecommunications services. The policy to permit private competition to international organizations built on complex political balances is, however, still questioned by many.

1.3. NATIONAL AND INTERNATIONAL DIFFERENCES

The need for telecommunications satellites arose, as we have seen, from the need for wider international communication which gave rise to the international partnership, INTELSAT. The future potential of satellites for improving national communications systems was also readily seen, especially by countries covering very large and uninhabited areas. It was also recognized by the founders of INTELSAT when they described the scope of the INTELSAT activities and included provision of domestic services in Article III on the same basis as international service for: b(i) countries with areas separated by high seas, or b(ii) countries with areas separated by land masses that make terrestrial systems unviable. The agreements came into effect on 12 February 1973 and the first request to use INTELSAT for domestic purposes came from Algeria only a few months later, closely followed by requests from the United States and others.

As examples of the two cases, the first country to use Article III b(i) was the United Sates using INTELSAT to link the mainland states with the state of Hawaii; the first country to use Article III b(ii) was Algeria, a large desert country in which terrestrial wideband facilities would be so expensive as to be prohibitive.

The agreements also recognized the likelihood for nations wishing to own their own telecommunications satellites to the extent that they do not cause radio frequency interference to INTELSAT, or other satellite systems, and that they do not cause significant economic harm to INTELSAT, which was created as a single global system. Each nation is, of course, solely responsible for its own internal communications systems as sovereign nations, having the right to regulate, and police the communications within its own borders. These sovereign rights of nations have always been jealously guarded by most nations. Most countries have a single telecommunications administration, usually government owned, which strongly opposes any communications networks that would bypass the nation's monopoly. Many countries organize their telecommunications so that profits from international services provide funds to finance basic telecommunications in other areas of their economy, and the reasons for this opposition are varied to the extent that their governments do not wish their communications to be influenced by foreign enterprises, or foreign governments, their state security undermined, or their revenues diminished.

Many countries have struggled hard to own their own national and international communications companies and facilities, and in almost all countries a large amount of government control is incorporated into any company involved in a nation's internal or external communications. Prior to the days of satellites the task of government control was comparatively easy as either cable beach heads or large antenna farms were needed to provide the services. The United States stood out alone as having a number of international record carriers competing for communications with other nations. These included such giants as AT&T, Western Union Telegraph Company (WUTCO), and RCA. Some European countries possessed companies engaged in international communications services such as France's France Cables et Radio, Italy's Italcable, and the United Kingdom's Cable and Wireless (C&W) which were partly or wholly government owned. All of these countries, including the United States, closely regulated the intrusion of foreign companies into their telecommunications bailiwicks.

The arrival of satellites has introduced the ability of one country to influence greatly many aspects of other countries and this possibility has been the subject of long and ongoing debates in the United Nations and other international groups. Discussions were first centered on the problem of imposition of international propaganda, broadcasting, television, major cultural changes, and similar concerns. These concerns remain and the technical ability for people to receive television programs directly from other countries still worries many nations. Many countries, especially the lesser developed ones, are very concerned at uncontrolled access to religious, educational, news, entertainment, and even pornographic programs emanating from the earth's more powerful nations.

A new dimension is now becoming practical resulting from more powerful satellites and much smaller earth stations. Satellite services are now being provided that permit the use of an earth station, transportable in two suitcases, to be used anywhere in the world, and these earth stations are likely to become much smaller in the years ahead. These micro earth stations were largely developed for military applications, mobile, and maritime use, but are now coming into use for private business applications, and receiving television from foreign countries. Many countries express concern that this type of earth station could be used clandestinely and not necessarily in the host country's interest, such as foreign exploration of national resources, commodity information, or even espionage activities.

These factors should be borne in mind when considering the many applications for satellite communications on an international basis, and why many countries support strong control and ownership of international telecommunications services.

During the first 20 years of satellite communications, a number of international and national networks have become established; they are described in later parts of this book. However, a new factor is now emerging—the advent of private competitive companies owning satellites for providing international services. These have come about since the U.S. Government permitted some privately owned companies to enter the field starting in 1984 under the umbrella of deregulatory policy.

The first such company, the Orion Satellite Corporation, was formed in 1982 by a former government telecommunications lawyer for the FCC, and received authority to plan to offer international satellite services in competition with INTELSAT. This authority, along with other similar applications, was granted by the FCC in 1985.

Within the United States this is a continuation of the deregularization process that started in the late 1960s when small companies emerged to compete with the national telephone giant, AT&T. These companies included Carterphone and MCI (originally called Microwave Communications Inc.). Like their predecessors, the newly emerging satellite system companies are interested only in the highly lucrative areas of the market—the high density traffic routes. The U.S. internal competition is for the highly lucrative long distance communications for business users, and long distance networks; the new satellite competitiors are mainly interested in the heavy transoceanic routes, or the North-South Americas region. None has yet shown interest in the less lucrative routes to most developing countries.

Thus far, 117 countries have signed the INTELSAT Agreements including the United States, and most other nations see the United States' actions as distinctly retrogressive, ignoring the fact that the U.S. deregulatory policies have permitted a wider variety of end-user services at a much lower cost. It is highly probable that in the years ahead many will look back at some of the U.S. actions regarding the deregulation of signatories as beneficial. The U.S. actions are likely to highlight the advantages of cooperative ownership, and the disadvantages of reliance upon small foreign companies.

The politics and economics of international versus national telecommunications are far from clear-cut, but some of the reasonings should at least be borne in mind by any business seeking to organize its international network, and they would be wise, at least in the longer term, to respect the aspirations and policies in countries where they wish to operate.

1.4. THE PRESENT STATUS

The enormous profitability of geosynchronous satellites and the lack of capacity existing for them in space were not foreseen by Arthur Clarke in 1945 or in the early 1960s when INTELSAT was being considered. The geostationary orbit, where satellites provide the earth with the fixed relay service which requires minimum or no tracking from the earth stations, is beginning to look crowded and is already becoming an important earth resource with serious political issues dividing further the have and have-not nations. With the increased interest in the geostationary orbit, satellite systems seem to be of four types:

Necessary systems
National/industrial reasons
National prestige
Paper satellites

All radio frequencies need to be registered with the ITU's International Radio Frequency Registration Board (IFRB) and this procedure applies also to frequencies used in the geosynchronous satellite orbit. As well as registering with the IFRB, member of countries of INTELSAT are obliged to coordinate any national satellite system with INTELSAT to ensure that their proposed satellite will not cause technical interference. If such systems would provide international service, they must also be coordinated with INTELSAT to protect INTELSAT from significant economic harm.

The frequency bands used for satellite communications are shown in Table 1.2; the satellites using the equatorial orbit, and the countries known to be planning satellite systems in late 1987, are provided in Appendix E.

The approval and methods for registering satellite frequencies has been, and continues to be, a major political debating point at the World Administrative Radio Conferences (WARC). It is not the purpose of this book to discuss the numerous issues involved, but there are some aspects of which the reader should be aware.

TABLE 1.2   Frequency Bands in Satellite Communications

Band	Frequency Bands From	To	Types of Service	
L	390 MHz –	1,550 MHz	MSS	Mobile satellite service
S	1.55 GHz –	5.2 GHz	FSS	Fixed satellite service
C	4.2 GHz –	6.2 GHz	BSS	Broadcasting satellites
K	10.9 GHz –	36.0 GHz	ISS	Intersatellite service
Q	36.0 GHz –	46.0 GHz		
V	46.0 GHz –	56.0 GHz		
W	56.0 GHz –	100.0 GHz		

Band	Range in MHz			Types of Service
L	399.9	–	400.05 Up	MSS, Radionavigation
L	406.0	–	406.1 Up	MSS/FSS
L	620.0	–	790.0 Down	BSS
L	1,215.0	–	1,240.0 Down	MSS, Radionavigation
L	1,240.0	–	1,260.0 Down	MSS, Radionavigation
L	1,530.0	–	1,533.0 Down	MSS, Land mobile
L	1,535.0	–	1,543.5 Down	MSS, Maritime
S	1,545.0	–	1,555.0 Down	MSS, Aeronautical
L	1,555.0	–	1,559.0 Down	MSS, Land mobile
S	1,559.0	–	1,626.5	MSS, Radiodetermination
S	1,631.0	–	1,634.5 Up	MSS, Land mobile
S	1,636.5	–	1,645.0 Up	MSS, Maritime
S	1,646.5	–	1,656.5 Up	MSS, Aeronautical
S	1,656.5	–	1,660.5 Up	MSS, Land mobile
S	2,500.0	–	2,690.0 Down	BSS
C	3,400.0	–	4,800.0 Down	FSS
C	5,000.0	–	5,250.0	ISS
C	5,850.0	–	7,075.0 Up	FSS
	7,250.0	–	7,750.0 Down	Government and military
	7,925.0	–	8,425.0 Up	Government and military
K	10,700.0	–	11,700.0 Down	FSS
K	11,700.0	–	12,500.0 Down	BSS
K	12,750.0	–	13,250.0 Up	FSS
K	14,000.0	–	14,500.0 Up	FSS
K	15,400.0	–	15,700.0	ISS
Ka	17,700.0	–	21,200.0 Down	FSS
Ka	22,550.0	–	23,555.0	ISS
Ka	27,500.0	–	31,000.0 Up	FSS
Ka	32,000.0	–	33,000.0	ISS
Q	40,000.0	–	41,000.0 Down	FSS or MSS
Q	40,500.0	–	42,500.0 Down	BSS
Q	43,500.0	–	47,000.0 Up	FSS or MSS
V	50,000.0	–	51,000.0	FSS
V	54,250.0	–	58,200.0	ISS
W	59,000.0	–	64,000.0	ISS

Notes: The C, Ku, and Ka frequency bands are shared with terrestrial communications services. These are only a guide, as several countries have exceptions, on a national or regional basis.

Some orbital locations are very much more desirable than others; this is evidenced by there being more applicants to use "slots" at these positions. The reasons are mainly geographical. If a satellite is for international use, maximum coverage is needed and positions over the middle of each ocean region provide the best worldwide coverage which is why INTELSAT originally chose 332.5°E to provide maximum coverage of Europe, Africa, and Asia from the eastern United States. Similarly, slots around 60°E provide widest coverage of Europe and the Far East, and around 175°E provides widest coverage of the Far East from the western edge of the United States.

If a satellite is required for national purposes only, it is preferable to use an orbital location as closely above the country as possible to minimize path transmission losses and maximize the country's geographic advantage. Considering the American continent, the equatorial orbit between approximately 60° and 140°W provides the best coverage.

Satellites have brought a whole new dimension to world communications, and allow human beings and businesses to communicate with each other more than had been dreamed possible only a few years ago. The mass distribution of television has permitted widespread distribution of news, sports, and entertainment programs, as well as a new dimension in educational possibilities that could permit the educational and progressive development of the world's poorest people.

By the end of 1986 there were already almost 100 telecommunications satellites transmitting to earth from the geosynchronous orbit and had there not been launch disasters, resulting launch delays, and problems in obtaining launch insurance, this number may well have been exceeded. Ownership of these satellites is diversifying quite quickly and within the next few years ownership may no longer be restricted to international organizations or even national governments, but open also to private or public companies. The overall control and legislation of this important earth resource is slowly, if not quickly, becoming the survival of the fittest. It is highly improbable that the ITU's IFRB will be able to control or police the activities of the geostationary arc. Although the INTELSAT system was intended to be a single global system, this organization has itself become almost a minor shareholder in the geostationary orbit having a total of 13 satellites in orbit to serve the entire world. This compares with over 50 commercial satellites owned by the United States (from *World Satellite Systems—1988 Directory*, prepared by the Communications Center of Clarksburg for Phillips Publishing Inc.), and an unknown number by the U.S.S.R.

One of the most uncertain factors affecting the present and the future of satellite communications is that of launching the satellites themselves. During the first 20 years of satellite communications, launching was an ever-decreasing risk; launch insurance had become possible, and many thought that satellite communications had become a low risk, high profit business. Then came a few launch failures by both the NASA shuttle and the European Ariane, and finally the tragic shuttle disaster in January 1986 which brought launching to all but a standstill for several years. These failures were costly, especially to the insurers whose rates are now almost prohibitive. I have no doubt in my own mind that the West's present launch problems will be overcome, but in the meantime and as long as Western satellite communications companies are considering the use of Soviet and Chinese launchers, however reliable they may prove, the launching and operation of satellites for communications will remain a high risk business. It is a business that still needs the financial and operational strength of the major international or regional organizations or the giant telecommunications companies.

It is clear that the U.S. telcommunications policy continues to play a very important role in global telecommunications satellites. It is equally clear that this policy, placed in the hands of a private corporation under the communications Satellite Act, has at times been a severe obstacle to the future of internationally owned satellite systems. During 1975, in a Study of Permanent Management Arrangements for INTELSAT, the consulting firm of Booz, Alen & Hamilton reported that "a majority of the Signatory representatives interviewed . . . . expressed concern over the conflicting pressures that COMSAT is exposed to in its multiple roles as:

Management services contractor to INTELSAT

The U.S. Signatory of INTELSAT (and later INMARSAT)

A U.S. commercial enterprise

A U.S. Government-authorized organization with the legal obligation to report annually to the Congress and the President

Given these multiple roles, it is inevitable that the Corporation will face potentially competing pressures, obligations and objectives." This situation was formulated in 1962 when the Communications Satellite Act was created but is much less of a problem today.

As a U.S. commercial enterprise over the years, COMSAT became "flush with profits from its U.S. monopoly over international satellite communications, and pushed into businesses that lost a total of US$200 million, including high-speed domestic satellite service, direct TV satellite broadcasts to homes, and manufacturing" (*Business Week Magazine*, 20 July 1987). These expansionary gambles never seemed to work and COMSAT's core international satellite service now faces growing competition.

As the U.S. Government-authorized organization to INTELSAT and INMARSAT, COMSAT still has multiple roles as part of the confusing turf wars of government agencies such as the FCC, National Telecommunications Information Agency (NTIA), and the Department of State. It is often difficult to observe a consistent U.S. policy towards international satellite telecommunications amidst the plethora of press coverage, but one thing seems certain—the policy to control the international organizations remains unchanged from that introduced by the Nixon Administration in 1969; this is despite the country's greatly diminished investment shares which have changed through greater world participation.

United States policy towards the international cooperative organizations is perhaps influenced by uninformed sources. One such policymaking guide was given by an eminent speaker at a satellite conference in Washington early in 1987. When describing the necessary actions by the U.S. delegation to the 1988 WARC-ORB conference suggested that some prime frequencies might need to be put into deep-freeze and that INTELSAT would have to learn to share the orbit. This type of disinformation must be seen for what it is when INTELSAT has 13 satellites in the geosynchronous orbit to serve some 172 countries, whereas the United States has over 50 satellites in commercial service alone, excluding military and other specialized satellites (*Satellite News*, 2 March 1987).

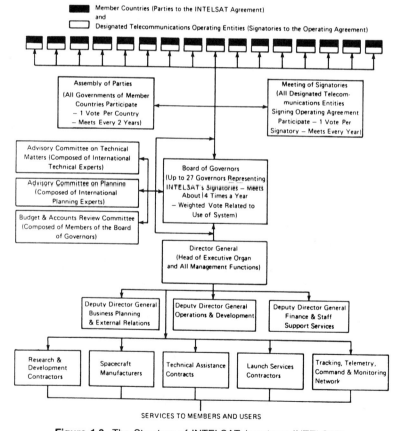

Figure 1.3. The Structure of INTELSAT (courtesy: INTELSAT).

2

INTERNATIONAL ORGANIZATIONS AS SERVICE PROVIDERS

Until 1988, the main providers of international satellite communications were international organizations in which countries grouped together to share resources and costs. This chapter describes each of these organizations and although some U.S. and Canadian domestic satellites operate limited transborder services between themselves and to neighboring countries covered by their satellite beams, these are not international organizations and are described later in Chapter 9.

2.1. INTELSAT

2.1.1. The Organization

INTELSAT was the first supplier of satellite communications services and has been supplying an ever-increasing number of countries with an ever-increasing number of services since it was first formed in 1964. At the end of 1988, INTELSAT was providing reliable, high quality international and domestic services to 165 countries by means of its 13 satellites. Of these 178 countries, 117 were members and were able to participate fully in the activities of the organization. These feats are not even approximated by other means of telecommunications.

Since INTELSAT first began supplying services in 1965, satellites have revolutionized societies round the globe by bringing live television coverage of major news and sports events to all countries of the world on a daily basis, as well as other low cost communication services. This all happened much faster than anyone envisaged at the time the United Nations put out its call for the peaceful exploration of outer space in 1961 (Resolution 1721(XVI)).

The success and growth of INTELSAT during its first 24 years has been largely the result of an efficient organizational structure, a solid financial base, and close continuing cooperation between the organization and its parties, signatories, and users. INTELSAT is an organization set up under an international treaty where each government signing the treaty is party to the agreement, and its representative is the signatory to the agreement. The signatory may be the Government's Ministry of Communications, or other government agency, or a company designated by the state to sign on behalf of the government. Countries may use INTELSAT services even if they are not members of it; that is, they have not signed the international agreements.

INTELSAT has enabled countries with different political systems and economic capabilities to collaborate in an efficient commercial telecommunications organization. INTELSAT has grown from a small consortium consisting of a small number of countries to a global organization where membership now comprises a majority of the states in the ITU; a country must be a member of the ITU to become a member of INTELSAT. The achievements of INTELSAT over its relatively short history demonstrate the enormously useful results that can be obtained by cooperative efforts between the nations of the world.

It is very important to understand a little about the aims and objectives of INTELSAT as well as some of the Agreements to which member countries have obligated themselves—and of course all companies and citizens of those countries.

Investment shares play an important role in INTELSAT because the organization is an international cooperative which means that a country's investment share is proportional to its measured use of the organization's services. Each country also receives appropriate returns for the use of its capital. The biggest users have the biggest investment and the most control on how their investment is used. Investment shares are calculated and adjusted annually on 1st March to reflect changes in the use of services. The minimum investment share that a country is permitted is 0.05 percent of the organization's valuation. At the start of 1988, the minimum investment share was around US$850,000. This valuation obviously increases most years taking into account new satellite launches, inflation, and many other items.

The financial arrangements of INTELSAT, being cooperative in nature, are that countries are repaid for the use of their capital at a rate of 14 percent per annum. This rate has existed since the organization was formed despite very considerable fluctuations in worldwide interest rates. This fact shows the conservative nature of the organization with regard to its financial security. INTELSAT accounts are published each year and an annual report can usually be obtained from its signatories or from the organization itself. A full list of INTELSAT signatories and their percentage investment share at 1 March 1988 is provided in Appendix A and the financial arrangements are shown pictorially in Figure 2.1.

INTELSAT's financial record has been quite remarkable and it is a perfect example of costs being reduced, and passing on these savings to the

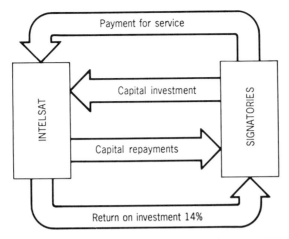

Figure 2.1. The INTELSAT financial arrangements (courtesy: INTELSAT).

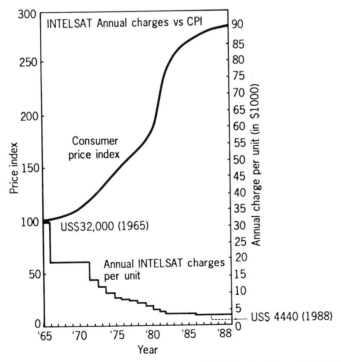

Figure 2.2. Comparison between charges and inflation (courtesy: INTELSAT).

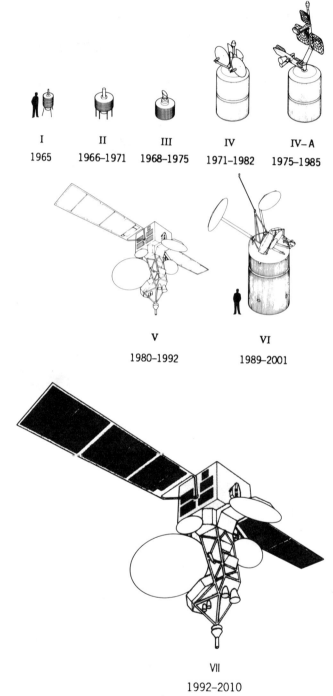

Figure 2.3. Evolution of INTELSAT satellites (courtesy: INTELSAT).

customer. It should be added, in fairness to INTELSAT, that INTELSAT's customers are its signatory owners and not the end-users of the services. INTELSAT charges have been progressively reduced despite the rise in inflation that has taken place and this is shown dramatically in Figure 2.2. Further reduction in INTELSAT's basic charge of US$740 per month per circuit is likely to have little or no impact on the end customer. Some examples of INTELSAT and end-user charges are given in later chapters.

The services supplied by INTELSAT to its signatories and nonmembers users come from a number of strategically located satellites in the geosynchronous orbit. Since the organization was first formed, the number and size of these satellites has grown continuously (Figure 2.3); major technological advances have been achieved with each new satellite series.

The satellites available in space must also include spare capacity to cover contingencies such as launch failure, satellite or component failure when in orbit, and the capability of accepting service growth. Once again, the advantage of INTELSAT's cooperative structure can be brought into use and the very considerable "spare" space segment capacity can be used to provide lower cost services to countries for domestic services. Over the years this spare capacity has given INTELSAT considerable extra revenue from leasing this capacity on a preemptive basis for domestic leased networks. Although these services were termed preemptive, it is worth noting that no domestic lease has ever been preempted. As the extent of this spare capacity has grown, there developed the economic need to use as much of the spare capacity as possible and making as much of it available as possible for revenue-producing services. The reliability and lifespan of satellites has increased considerably but even holding one large satellite in space as a spare has become a very expensive luxury. As a consequence much wider use is now being made of this spare capacity to minimize all tariffs, to offer more and more attractive services at competitive costs, and also to make most efficient use of the earth's geostationary arc resources.

While established procedures within INTELSAT make cost recovery an objective as far as is practical, the agreements do not require strict cost-based pricing, which permits flexibility in establishing rates.

All users of the INTELSAT system are charged the same rate for the same services. Rates are averaged, based on total use of the entire global system and reflect the costs associated with providing the services.

2.1.2. INTELSAT Satellites

Since INTELSAT's operation started in 1965, traffic growth has been so rapid that invariably the capacity available in any individual satellite has been quickly filled up resulting in the need for larger satellites with increased capacity. This situation is due to the fact that for maximum international connectivity, all countries would like to access one satellite, which would require one gigantic satellite in each ocean region.

TABLE 2.1   Orbital Status of INTELSAT Satellites—1988

Satellite	Primary Operational Role	Location
	Atlantic Ocean Region	
INTELSAT V-A (F-13)	INTELSAT business service and domestic lease services	307°E
INTELSAT V (F-4)	International and domestic lease services	325.5°E
INTELSAT V-A (F-11)	International and domestic lease services	332.5°E
INTELSAT V-A (F-10)	International services	335.5°E
INTELSAT IV-A (F-4)	Domestic lease service	338.5°E
INTELSAT V (F-6)	International and maritime services	342°E
INTELSAT V (F-2)	International and domestic lease services	359°E
	Indian Ocean Region	
INTELSAT V-A (F-12)	International services	60°E
INTELSAT V (F-5)	International, domestic and maritime services	63°E
INTELSAT V (F-7)	Domestic lease services and maritime services	66°E
	Pacific Ocean Region	
INTELSAT V (F-3)	International services	174°E
INTELSAT IV-A (F-3)	Spare in inclined orbit	177°E
INTELSAT V (F-1)	Domestic lease services	177°E
INTELSAT V (F-8)	International and maritime services	180°E

Each successive satellite series needed to develop new techniques to obtain the increased capacity needed within the bounds of current technology and the limitations of satellite design such as weight and power available. A brief description follows of each type of satellite developed by INTELSAT for its global system and the orbital status in 1988 is shown in Table 2.1.

INTELSAT I (Early Bird). The first satellite launched by the new organization was based upon the earlier SYNCOM satellites developed by the Hughes Aircraft Company of the United States. It was more commonly known at the time as Early Bird, although it was also known by the manufacturer's designation HS 303.

Early Bird had two transponders, each having about 30 MHz bandwidth, and a transmit power of about 4 W. Its antenna system had twice the gain of its SYNCOM predecessors and illuminated only the North Atlantic region with an effective isotropic radiated power (EIRP) of about 12 to 14 dBW. It was a single spin stabilized type where the satellite is spun about the axis on which the moment of inertia is maximum, so that the axis is maintained constant in space by the inertial system of the gyro effect. The spinning body must be equipped with a nutation damper to absorb the disturbing torque.

The total available prime power, provided by about 6000 solar cells around the satellite's cylindrical surface, was 45 W. The Early Bird operating frequencies conformed with the new ITU allocations at that time for fixed satellite services, receiving on the up-link in the 6 GHz band and transmitting on the down-link in the 4 GHz band. INTELSAT I, although not spherical, was about 0.7 m in diameter, 0.6 m high, and weighed 68 kg at launch. It was placed into a geosynchronous orbit 35,800 km above the equator over the middle of the Atlantic ocean in April 1965. After many weeks of testing it went into service in June of the same year, providing initially about 60 circuits between Andover, Maine in the United States, and two earth stations in Europe (Goonhilly Downs in Cornwall, England, and Pleumeur Bodou in north west France).

Later in 1965, two more stations had been built in Europe, one at Raisting in Germany, and the other at Fucino in Italy. However, the characteristics of the satellite were such that only two earth stations could access the satellite at the same time, one through each transponder, and therefore it was possible to establish only one point-to-point trunk between Europe and the United States. To enable all the European earth stations to participate in this venture, a three-week rota was established between the German, French, and British earth stations such that one station would be operational for a week while a second was on standby and the third was available for maintenance and modification purposes. This rota was effective for five days a week, leaving the Italian earth station to provide the service over weekends with one of the others acting as standby. It was during this early period that satellite communications were really made visible to the general public; live television exchanges could now take place across the North Atlantic.

On this basis, operations grew over the next two years from about 60 circuits per 12-hour day for five days a week, to about 240 circuits per 24-hour day. Also during that time, an earth station at Mill Village, Nova Scotia, Canada joined the group and shared time with the U.S. Andover earth station. The satellite could be used either for telephony service or for one good-quality television program. INTELSAT I was expected to have a lifespan of 18 months, but provided regular transatlantic service for over three years, and was still usable five years after its launch.

INTELSAT II. The first INTELSAT II satellite was a launch failure, but the second launch, four months later in January 1967, was successful. This series was an improved version of the HS 303, and was designated by the manufacturer as the HS 303A. Each satellite was expected to last three years and had almost twice the diameter and weight of the earlier version. INTELSAT II had a 120 MHz transponder design permitting multiple-access operation. The output power was also greater with a 70 W power source, and a battery to provide power during eclipse periods. The antenna system had a wider beam and covered both northern and southern hemispheres thus taking satellite communications further towards global coverage. The output power of the satellite was 12 to 14 dBW. Although the capacity remained about the same at 240 circuits, and because of its multiple-access capability, it was now possible to provide telephone and television services simultaneously.

The new satellite enabled all the earth stations in the region (seven at the time) to operate continuously rather than on a rota basis, and before the end of 1967 another satellite was launched over the Pacific ocean extending the satellite system to cover two of the three ocean regions of the world. It was this series that provided so much support to the U.S. Apollo mission network culminating with man's arrival on the Moon in 1969—televised worldwide by INTELSAT using its INTELSAT I and II series satellites.

INTELSAT III. Apart from being bigger and better, the INTELSAT III series had a new polarization system and was supplied by TRW Inc. (Thompson, Raymo, and Wallridge). Earlier satellites had used linearly polarized waves for sending the magnetic radio waves through space; INTELSAT III used circularly polarized waves. This design change was to avoid Faraday rotation effects and antenna beam polarization alignment requirements. For singly polarized linear systems such as INTELSAT I and II, these effects impacted mainly on the power loss of satellite point-over capability.

INTELSAT needed to serve a large number of earth stations, some with the low elevation angles, which required spacecraft antennas providing multiple global beams. The question of circular versus linear polarization at C-band must be considered from an overall system viewpoint and the disadvantages of dual linear polarization for INTELSAT are briefly:

It complicates the spacecraft antenna design if multiple-beam frequency-reuse beam coverages are required; this is necessary for maximum use of the available orbital resources

It complicates earth station point-over to spare, or other satellite locations. The polarization alignment of the earth station must be adjusted periodically; this is not necessary in dual circular systems.

Use of half wave rotatable polarizer systems necessary for dual linear systems have at least as much cost impact on small terminals as the quarter wavelength polarizer requirements for dual circular systems.

Faraday rotation effects become a factor for dual linear systems; this can be just ignored or periodically adjusted.

INTELSAT, through COMSAT, which was managing INTELSAT at the time, chose the path of circular polarization for the INTELSAT III and subsequent series. Many other satellite systems have chosen linear polarization at C-band. This factor was to become of some importance later with regard to domestic systems and their compatability with the INTELSAT system.

Once again the first of the new series was a launch failure, but by the end of 1968 the second launch was successful and the satellite was placed into service over the Pacific. The third launch placed a second satellite over the Pacific and the original INTELSAT III was transferred to the Indian ocean region thus establishing a truly global satellite system by July 1969.

The INTELSAT III satellites were 1.4 m in diameter and 1 m high and weighed nearly 300 kg at launch. Their prime power had increased to 120 W including the battery. The antenna system was "despun," capable of pointing the high-gain antenna beam toward the earth by giving the antenna spin at the same speed as the satellite spin but in the opposite direction.

INTELSAT III used both 500 MHz bands allocated for the fixed-service satellite communications in the 4 and 6 GHz frequency bands (C-band), and had two transponders, each of 225 MHz bandwidth. Global beam antennas were used to provide full earth coverage with an EIRP of 22 dBW and a total capacity of about 1200 circuits plus one television channel.

The number of earth stations was increasing rapidly and it was soon necessary to introduce a two-satellite system over the Atlantic, although a single satellite remained sufficient in both the Pacific and Indian ocean regions. The assignment of earth stations between the two Atlantic satellites was done on a community of interest basis; one providing service between the United States, Canada, France, United Kingdom, and the countries of the Middle East and Africa, while the other provided service between the United States, Germany, Italy, Spain, and the countries of South America.

By the end of 1970, there were 20 earth solutions operating in the Atlantic ocean region, 14 in the Pacific, and 12 in the Indian ocean region representing an increase of about 30 earth stations in two years.

Despite this growth, the INTELSAT III series was not altogether successful. Of the eight attempted launches, three were failures, two had serious antenna despin problems resulting in their being retired from service earlier than had been anticipated, and two others had communications equipment problems resulting in a loss of circuit capacity. Although the design-life was five years, the maximum obtained was barely three and a half years, and backup facilities for each ocean region were provided by the obsolete INTELSAT II satellites.

INTELSAT IV. This series was an extremely successful series for INTELSAT and similar satellites also became the basic workhorse for many domestic satellite systems.

The first INTELSAT IV, again supplied by Hughes Aircraft, was launched for the Atlantic region in January 1971 and started operations in March. The second was launched nearly a year later, also for the Atlantic region. Shortly afterwards, other satellites were launched for the Pacific and Indian ocean regions, establishing a full global INTELSAT IV system during 1973. Only the sixth launch was a failure, and the success of the

INTELSAT IV series contrasted greatly with the earlier INTELSAT III performance.

The INTELSAT IV satellites were about 2.4 m in diameter, 2.8 m high (5.3 m including the antenna structure) and weighed almost 1400 kg. They used the same two 500 MHz frequency bands but by using 12 separate transponders, the spectrum was divided into 36 MHz segments which became almost a standard transponder bandwidth.

The prime power supply, developed from some 42,000 solar cells, was rated for approximately 500 W at 24 V and included two 15 ampere-hour batteries to ensure continuity of operations during eclipse periods. All 12 transponders received from the earth stations via a global-beam antenna system similar to the earlier INTELSAT II and III satellites. This satellite was another type of dual spin satellite called a despun platform type. As the number of antennas mounted on a satellite increases, the feeding of the radio frequency power to the antennas through the rotary joint becomes much more difficult. In order to overcome this difficulty, the antennas and the transponders were mounted on a platform that is mechanically despun as one body. In this type, the electric power is supplied from the power source mounted on the spinning body through slip rings and the control signals are connected via rotary transformers. The connecting unit is called the bearing and power transfer assembly (BAPTA). The INTELSAT IV, IV-A, and VI series all have this type of design. In the despun platform type, the platform despins about the axis on which the moment of inertia is minimum, and therefore, the conditions of gyrostat stabilization, which is different from that of the single spin stabilization, must be satisfied. To attain this condition, a nutation damper which is more effective than that on the spinning body is mounted on the despun portion.

In addition to the global-beam transmit antenna system, which has a beamwidth of 17°, the INTELSAT IVs had two narrow-beam, or spot-beam, transmit antennas of 4.5° beam width, which could be steered from the ground to illuminate two specific areas of the earth's surface. Of the 12 transponders, four could be switched to the global beam or one of the spot-beam antennas; another four to the global beam of the other spot-beam antenna and the remaining four were connected permanently to the global-beam antenna. The two spot-beam antennas were normally deployed towards the east and west of an ocean region, and hence the associated transponders were referred to as the east spot and west spot transponders, respectively. The INTELSAT IV global-beam transmissions were similar in power to those of the INTELSAT III with an EIRP of 22 dBW, while the spot beams had an EIRP of 33.7 dBW. The objective of the spot beams was to provide more capacity for the heavy east-west traffic requirements in each ocean region as the spot-beam carriers could provide approximately double the circuit capacity of similar sized global-beam carriers. This design by INTELSAT was also found to be particularly useful for providing higher power for domestic satellite systems where the increased power could be used for designing a ground system with much smaller earth stations.

INTELSAT IV-A. With the rapid growth of traffic, the capacity offered by the INTELSAT IV satellites quickly became inadequate and an interim satellite was again needed until the next generation became available. The INTELSAT IV was modified to provide more spot beams by creating spatial separation between the eastern and western beams and thus reusing the frequency spectrum and almost doubling the capacity of the satellite.

Through this frequency reuse, the INTELSAT IV-A had a total of 20 transponders compared with the 12 of the INTELSAT IV, and by using the extra EIRP available through further spot-beaming, channel capacity was increased to around 6000 channels plus two television channels.

Since the actual shape of the antenna beam coverage area was controlled by the dish antenna feed, which was made up of a bank waveguide radiators, beam shaping and positioning could be achieved by appropriate phasing between the radiating elements.

The switching arrangements for the various amplifiers within the satellite was more complex but in all other respects the INTELSAT IV-A was the same as the INTELSAT IV with regard to lifespan and power arrangements.

INTELSAT V. In the late 1970s traffic requirements were still growing faster than had been expected and the new INTELSAT V series was eagerly awaited just had been the arrival of earlier series. The new series of satellites each had a capacity of around 11,000 circuits plus television and in 1987 the INTELSAT V continues to be the mainstay of the global system.

The INTELSAT V series, shown in Figure 2.4, was manufactured by Ford Aerospace and Communications Corporation (FACC) and incorporated many design features that were different from earlier satellites. In addition, of great political significance to an international organization such as INTELSAT, many other countries apart from the United States contributed to the design and manufacture including:

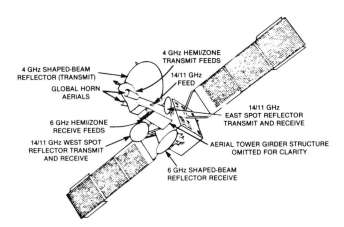

Figure 2.4. INTELSAT V (courtesy: INTELSAT).

Aerospatiale (France)

GEC-Marconi (United Kingdom)

Messerschmitt–Bolkow–Blohm (West Germany)

Mitsubishi Electric Corporation (Japan)

Selenia (Italy)

Thomson-CSF (France)

Many countries now have their own space programs and associated industries and consequently favor international enterprises that manufacture large satellite series such as the INTELSAT V of which 15 were finally ordered. Each of the above manufacturers concentrated on specific subsystems.

Aerospace initiated the structural design that forms the main part of the spacecraft modular design construction, and supplied the main body structure thermal analysis and control. GEC-Marconi manufactured the 11 GHz beacon transmitter for earth station tracking. Mitsubishi was responsible for both the 6 GHz and 4 GHz global antennas, and also manufactured the power control electronics. Using an FACC design, it also manufactured the telemetry and command digital units. Selenia built the six telemetry, command and ranging antennas, two 11 GHz beacon antennas and two 14/11 GHz spot-beam antennas. It also built the command receiver and telemetry transmitter, which combines to form a ranging transponder to determine the spacecraft position in the transfer orbit. Thomson-CSF built the 10 W, 11 GHz traveling wave tubes of which there are 10 in each satellite. FACC is responsible for the development of the communications packages, including the Maritime Communications Package (MCS) built into F-6, 7, 8, and 9 satellites and the business packages built into the later IS-VA models.

Some of the technological features used in the INTELSAT V include the use of Ku-band and further frequency reuse through polarization isolation techniques in the C-band frequencies. The bandwidth of the Ku-band is nearly doubled using beam separation techniques. Linear polarization is used for the Ku-band beams with orthogonal polarizations for the up- and down-link transmissions. In the C-band, double frequency reuse of part of the frequency band is obtained using beam separation techniques similar to those used in the INTELSAT IV-A satellites. These frequency bands can be used for west-to-east, east-to-west, west-to-west, or east-to-east transmissions. In addition, a further double frequency reuse is obtained by using orthogonally polarized transmission, which together with the beam separation provides fourfold frequency reuse of a part of the C-band.

A new design principle called three-axis stabilization was also used in the INTELSAT V whereby the body of the satellite remains stationary and a spinning gyro keeps the satellite, with its fixed antenna, pointing towards earth. With this technique there is no complicated despinning machinery

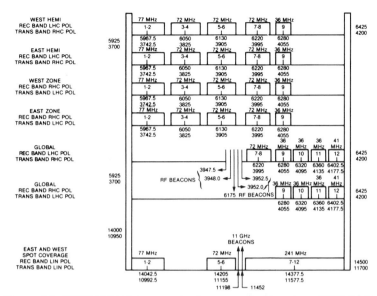

Figure 2.5. INTELSAT V-A transponder frequency plan (courtesy: INTELSAT).

and the satellite can be lighter to launch allowing more solar cells to be mounted on the solar arrays.

Two separate feed arrays radiate the C-band crosspolarized hemispheric and zone beams. The Ku-band spot beams are generated by horn antennas. The transponders are configured with 15 receivers using gallium arsenide field effect transistors (GaAs FET) preamplifiers, seven input multiplexers of graphite epoxy construction, a matrix switching exchange with 35 switches, up-conversion to 11 GHz for 10 channels, followed by 43 TWT amplifiers in a redundant configuration. The 4 GHz output multiplexer is of the continuous band design which eliminates the conventional odd/even channel multiplexers and allows a single transmit antenna and a single receive antenna to perform the hemispheric/zone function which would otherwise need two transmit antennas and one receive antenna.

The communications subsystem features 27 simultaneous, independent channels with bandwidths ranging from 34 to 241 MHz per transponder as shown in Figure 2.5. Flexibility in the interconnection through the switching exchange results in 534 combiations of global, hemispheric, zone, and spot beam coverage. The total channel bandwidth of 2241 MHz is achieved in an RF spectrum of only 912 MHz using these techniques. Some of the most salient features of the INTELSAT V are:

General Characteristics

Three axis stabilization with sun and earth sensors

Solars arrays producing 1290 W of power

Modular construction
Seven year design life

Communications Characteristics

12,000 circuit capacity plus two TV channels
C- and Ku-bands
Fourfold frequency reuse at C-band
Double frequency reuse at Ku-band

Spacecraft

Aluminum main body structure
Graphite epoxy antenna tower
Catalytical and electrothermal hydrazine thrusters

INTELSAT V was the first satellite series to use the long-planned time division multiple access (TDMA) digital modulation technique. The development of this digital modulation technique had been recommended by COMSAT in its role as interim manager in the mid-1960s and much of the 20 years research and development of this took place at COMSAT's own laboratories. During the 1970s COMSAT had wished to introduce this system operationally but had been dissuaded from doing so until some degree of reliability could be introduced. The 20 years of research and development proved extremely expensive to INTELSAT and its signatories, but has permitted a major step towards digital transmission techniques. The new terminal equipment has been also extremely expensive for the signatories that had to install it. Original plans were to introduce TDMA on high density routes in the North Atlantic but due to changes in traffic patterns most of the TDMA traffic terminals installed thus far have been in the Indian ocean region. Although the research and development was expensive, it did lead the way towards a global digital system.

INTELSAT V-A. Even before the first launch of the INTELSAT V series had begun, it was obvious that more capacity would be needed in the INTELSAT system by the mid 1980s so plans were again started for a modified INTELSAT V, called the INTELSAT V-A.

Preliminary planning was started by FACC in the 1980 with the objective of having the satellites available in 1984 (the first INTELSAT V-A was actually launched in 1985). One of the main new features was the addition of two steerable C-band spot-beam antennas, and the last three spacecraft have modifications to provide customer-premise earth stations used with INTELSAT Business Services (IBS). These features are:

Operation of Ku-band frequencies specifically suited for IBS

Improved connectivity between Ku-band and C-band to permit global access for IBS services

Larger west spot-beam coverage

INTELSAT VI. INTELSAT VI, shown in Figure 2.6, when launched in 1989, will be the largest commercial communications satellite program ever undertaken. The contract was awarded to Hughes Aircraft Corporation following an international competitive tender.

The engineering development and production of the first five spacecraft was contracted by INTELSAT on a firm, fixed price basis for US$700 million. The same contract contains options for 11 more spacecraft in the INTELSAT VI series.

Each INTELSAT VI satellite will have a capacity of 33,000 4-kHz voice circuits, or about 100,000 circuits using new compression techniques—three times the capacity of the INTELSAT V/V-A—and have twice the payload mass. In orbit, each will weigh almost 1800 kg, be 3.6 m in diameter, and when deployed will stand 11.7 m. The solar panels will generate 2250 W of power for the satellite.

The INTELSAT VI design combines important features of two satellites now in production by Hughes. The first is the HS 376 extended power spinner, which has become popular for domestic communications satellite systems. The second is the widebody spacecraft originally designed for launch on the Space Transportation System (shuttle), but may also be

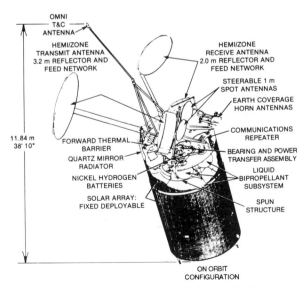

Figure 2.6. INTELSAT VI (courtesy: INTELSAT).

launched on the Ariane or some other launch vehicles. This alternative proved to be fortuitous following the shuttle disaster and an alternative launcher was needed, the Titan rocket.

INTELSAT VI produces substantially more power than earlier Hughes satellites by using two large, concentric cylindrical solar arrays. During launch it is very similar to a telescope as the outer array is pulled over the inner array and the main body. Once in orbit, the outer array is extended downward to expose the inner panel, greatly increasing the available area for the solar cells. Other features of the HS 376 include radial heat rejection, system dynamics and stabilization, and antennas that are folded during the launch. The original design also included the "frisbee" method of launching from the shuttle in which a simple spring and latch mechanism gives an upward motion to the satellite causing it to leave the launch vehicle. This also provides a slow spin for initial gyroscopic stability.

The communications subsystem provides 46 distinct transponders and both C- and Ku-band frequencies. The C-band payload includes two hemispheric repeaters, four zone-beam repeaters, and two global-beam repeaters. The Ku-band payload consists of two spot-beam repeaters. A block diagram of the satellite communications subsystem is shown in Figure 2.7. The receiver sections contain a total of 20 receivers arranged in five groups of 4-for-2 redundancy which serve all the repeaters. Solid state components are used throughout.

The power amplifier sections contain driver amplifiers, upconverters, traveling wave tube amplifiers (TWTAs), and solid state power amplifiers (SSPAs) in various combinations. In the 4 GHz channels, TSTAs with power levels from 5.5 to 16.0 W are used for the hemi, large zone, and global

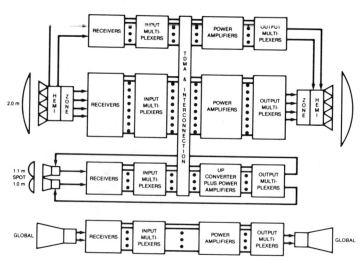

Figure 2.7. INTELSAT VI communications subsystem (simplified) (courtesy: INTELSAT).

repeaters. The small zone repeaters use SSPAs because of their low power requirement. The SSPAs use highly reliable gallium arsenide field effect transistors.

Ku-band TWTAs, used in the INTELSAT V, have been modified from 10 to 8.5 W for the spot repeater at 11 GHz. Power amplifiers are 3-for-2 redundant in all C-band channels and 4-for-2 redundant in the Ku-band channels. The output multiplexers for all repeaters use continuous channel multiplexing with invar filters. The transponders are interconnectible using either static switch matrices or a new dynamic switching network which provides satellite switched time division multiple access (SS-TDMA) capability. Static interconnection matrices are provided for all channels except the global repeaters. Channels 1-2 and 3-4 are connected by dynamic switches to provide SS-TDMA operation. Each of these switching units consists of a dynamic microwave switch matrix, with its associated digital control logic, and an input ring redundancy network with a bypass switch matrix. The dynamic switch is operated by redundant distribution and control units, each with three memories and a very stable timing source. This switch for INTELSAT VI is an important new technology. The transponder frequency plan is shown in Figure 2.8. Different types of beams

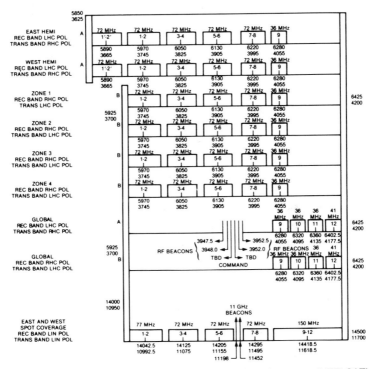

Figure 2.8. INTELSAT VI transponder frequency plan (courtesy: INTELSAT).

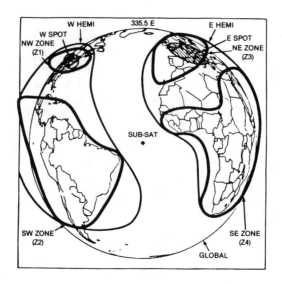

NOTE:
1. THE EAST 14/11 GHz SPOT BEAM IS STEERABLE OVER THE FULL EARTH'S DISC.
2. THE WEST 14/11 GHz SPOT BEAM IS STEERABLE OVER THE WESTERN HEMISPHERE ONLY.

Figure 2.9. INTELSAT VI Atlantic ocean coverage from 335.5°E (courtesy: INTELSAT).

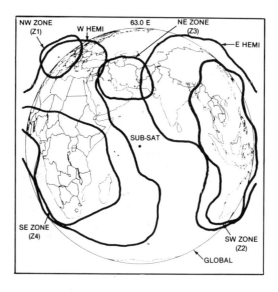

NOTE:
SEE FIGURE AT 335.5 E FOR 14/11 GHz EAST AND WEST SPOT BEAM CAPABILITIES.

Figure 2.10. INTELSAT VI Indian ocean coverage from 63°E (courtesy: INTELSAT).

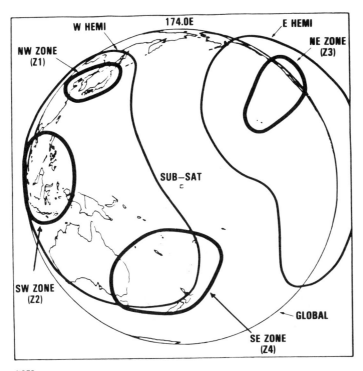

NOTE
SEE FIGURE AT 335.5 E FOR 14/11 GHz EAST AND WEST SPOT BEAM CAPABILITIES.

Figure 2.11. INTELSAT VI Pacific ocean coverage from 174°E (courtesy: INTELSAT).

can be transmitted and received by the INTELSAT VI satellites as shown in Figures 2.9, 2.10, and 2.11 which show the approximate coverages for the three main ocean regions.

INTELSAT VII. INTELSAT went to international tender for the next series of satellites (INTELSAT VII) in late 1987 and received four proposals early in 1988 from:

1. Ford Aerospace which had formed a group together with France's Alcatel, and Japan's Mitsubishi
2. General Electric Astrospace which had formed a group together with Germany's Messerschmitt–Bolkow–Blohm (MBB) and France's Aerospatiale
3. General Motors' Hughes Aircraft Corporation
4. A multinational group lead by France's MATRA that included TRW, British Aerospace, Canada's SPAR, Italy's Selenio Spazio, the Netherlands' Fokker, and Germany's ANT.

The proposal from Ford Aerospace was selected for further negotiations, and later contracted in October 1988 for a minimum of five INTELSAT VII satellites with options for further purchases at any time through 1994 at agreed option prices.

The first of the INTELSAT VII series is scheduled for launch in mid-1992, and the second in early 1993 for use over the Pacific ocean, replacing INTELSAT V-A satellites which will be at the end of their planned lifetime. Later launches will place INTELSAT VII over the other ocean regions. They will be particularly attractive to users round the Pacific rim as they will be considerably more versatile than the present INTELSAT V and will have the ability to switch C-band capacity between east and west hemisphere using an independently steerable spot beam. A third K-band antenna with transponder switching will also greatly improve the K-band flexibility and extend it to new areas.

INTELSAT VII will, in appearance, be similar to the INTELSAT V-A but will have more power and be more versatile. The improvements are expected to encourage a growth in smaller earth stations that will provide new telecommunications services to many communities and businesses. Each INTELSAT VII will have 36 transponders: sixteen 72-MHz and ten 36-MHz C-band transponders, and six 72-MHz and four 112-MHz Ku-band transponders.

2.1.3.   INTELSAT Earth Stations

In order to obtain operational conformity within the system, a number of standards for earth station performances and services have evolved. The full performance characteristics are available from INTELSAT through its signatories.

Standard A. This standard generally calls for a large parabolic antenna (typically 15 to 30 m in diameter) and is the most widely used earth station in the system for large scale international telecommunications. It's figure-of-merit (G/T) used to be 40.7 dB/K but was reduced in 1985 to 35.0 dB/K to permit smaller earth stations, and operates in the 6/4 GHz frequency bands. The Standard A stations allow the most efficient use of the satellite orbit and the frequency bands allocated to the fixed satellite service (FSS).

Standard B. This standard was developed for situations that called for a more economical alternative to the Standard A. A Standard B earth station, which usually has a parabolic antenna of about 11 m diameter and operates in the 6/4 frequency bands, is particulary suitable for countries with small traffic requirements. Its figure-of-merit (G/T) is nominally 31.7 dB/K. Some countries use a Standard B earth station when using INTELSAT for the first time, later upgrading to a Standard A when traffic growth warrants a larger investment, or a Standard B earth station may be used as an emergency backup earth station for other facilities.

Standard C. This earth station has an antenna usually between 14 and 19 m in diameter, and is designed specifically for use with the 14/11 GHz frequency bands. Its figure-of-merit (G/T) is nominally about 40 dB/K.

Standard D. This type of earth station was introduced for the low-density VISTA service and has two varieties. The Standard D-1 is a small antenna of about 5 m diameter for rural or remote areas, and provides minimal capacity at the lowest possible cost. The Standard D-2 is similar in size to a Standard B earth station (about 11 m in diameter) and is used as the central earth station in a VISTA network.

Standard E. These three types of earth stations operate in the Ku-band 14/11 GHz frequency bands and are used with the INTELSAT Business Service (IBS). They are designated the Standard E-1 (3.5 m), E-2 (5.5 m), and E-3 (8 to 10 m), and have G/Ts of 25 dB/K, 29 dB/K, and 34 dB/K, respectively.

Standard F. These three types of earth station are similar to the Standard E except that they operate in the C-band 6/4 GHz frequency band, and are designated F-1 (5 m), F-2 (7 m), and F-3 (9 m).

Standard G. These earth stations can access INTELSAT satellites through the use of fractional transponder power and bandwidth lease definitions or the use of other space segment facilities such as INTELNET services.

Standard Z. These are domestic earth stations that use either C-band or K-band frequencies. They have no specified figure-of-merit (G/T), antenna size or modulation method.

Standard G and Z earth stations differ from other standards in that many features are not specified including:

Maximum EIRP per carrier
Modulation method
G/T
Transmit gain
Channel quality

The definition of these parameters is essentially left to the user in order that it may determine the best transmission design for the particular requirements. The actual values selected for these parameters within the overall transmission plan must, however, be reviewed and approved by INTELSAT for the benefit of other system users.

If the requirements for Standard G or Z performance are met, a wide range of earth station sizes, modulation techniques, and performance qualities can be chosen by individual countries.

Non-Standard Earth Stations. There are a few non-standard earth stations operating within the INTELSAT system but these are mainly special purpose or experimental earth stations approved individually by the INTELSAT Board of Governors.

2.1.4.   Summary of Services

INTELSAT provides telecommunications capacity to the telecommunications authorities in the different countries and they, in turn, use this capacity to provide telephone calls, leased circuits, telegraph, data, and television services to their customers. Almost anything that can be classified as telecommunications can be carried by INTELSAT, and new services are constantly being considered to meet the requirements of the end-users. Figure 2.12 shows the types of services provided and the percentage of the total revenue that each contributes to INTELSAT. A full description of INTELSAT services, terms and conditions, and tariffs can usually be obtained from INTELSAT or its signatories in any respective country, but it must be remembered that INTELSAT tariffs do not include the signatory content of the service to the end-user—the INTELSAT tariff is only for the space section of the overall link.

INTELSAT generally leases satellite capacity in half circuits, each of which can be described as the amount of capacity required to provide one end of a two-way telephony circuit which, when matched with a half circuit at the other end, provides a full circuit between two earth stations. Two such half circuits are required for each telephone conversation or two-way data circuit. Multiples of this basic unit can be used for services on a full-time, part-time, or occasional use basis for:

Telephony
Telegraphy and other record services
A wide range of digital services
Television
Television and videoconferencing
Cable and emergency restoration
Radio broadcasts

For full time preassigned telephone, telegraph, and data services, INTELSAT leases satellite capacity to telecommunications authorities on the basis of a "unit of utilization" which is defined as the measure of entitlement, secured through the allotment by INTELSAT of space segment capacity, for the establishment of one end of a two-way 4-KHz telephone circuit providing, as an objective, quality of service in accordance with appropriate CCITT/CCIR recommendations, by means of:

Intelsat Revenues 1987
(Estimated percent)

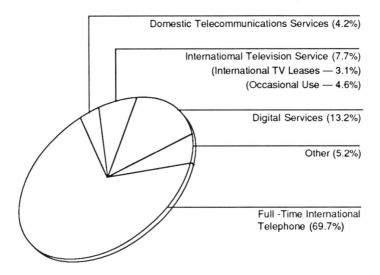

Domestic Telecommunications Services (4.2%)

Internatiomal Television Service (7.7%)
(International TV Leases — 3.1%)
(Occasional Use — 4.6%)

Digital Services (13.2%)

Other (5.2%)

Full -Time International
Telephone (69.7%)

Disposition of Total Revenues
(Millions U.S.$)

	Compensation For Use Of Capital	Repayment Of Capital	Operating Expenses	Total Revenues
1977	69	58	30	157
1978	81	70	35	186
1979	96	63	36	195
1980	114	56	46	216
1981	118	75	56	249
1982	160	89	66	345
1983	189	101	76	366
1984	208	115	88	411
1985	224	129	104	457
1986	207	163	118	488
1987	196	213	110	519

Figure 2.12. INTELSAT revenues and services (courtesy: INTELSAT).

a. Access to a satellite in the multichannel mode
b. An approved earth station conforming to the requirements set forth in INTELSAT documents and all amendments thereto.

Such a circuit may also be used for voice frequency telegraphy, facsimile telegraphy, data, or radio broadcast signals. Digital circuit multiplication techniques are also used to make the satellite use more efficient.

Preassigned single channel per carrier (SCPC/PCM/PSK) services may be used also, but these are usually used to and from Standard B earth stations for telegraphy and telephony services. They may also be used for one-way or two-way multidestination data services.

Demand assigned services are also available on one Atlantic ocean region satellite through the SPADE transponder. SPADE (single channel per carrier PCM multiple access demand assignment equipment) is provided for those countries equipped with this type of equipment and is particularly suited to those routes where traffic does not warrant the provision of full-time circuits. The charge for this service is based on the per minute of satellite circuit holding time and is charged to each end of the circuit.

For scheduled television services, video channels are available on each INTELSAT satellite for television services on a scheduled basis using 20 MHz video carriers with subcarrier audio. Television may be single or multidestination and double-hop transmissions over two ocean regions are both possible and frequent applications. Television is noted for the complex interface arrangement necessary for the multidestination transmissions, particularly with video networks having different transmission standards. Orders for television programs are collected by the international tele-communications carriers in the various countries and passed onto the INTELSAT Operations Center in Washington D.C. where orders are coordinated by the TV Service Center.

In 1981, an international leased television channel service was introduced for full-time television services between two or more countries allowing the lease of capacity for television service in units of 18 MHz. This service proved to be very popular and was expanded considerably during 1984 to cover a wider range of user requirements. These services are now provided in the most economic manner on any available satellite, short-term or long-term leases, with a number of service priorities and corresponding tariffs.

Video services are also possible. There are two distinct services that are commonly referred to as "teleconferencing." The first is a seminar type conference with a professional moderator and usually consists of one-way video together with audio feedback. This form of teleconferencing is most appropriate for educational and instructural purposes. It has been used to project annual corporate meetings to a number of widespread locations, where shareholders have an opportunity to view the proceedings on a large screen video display and to question the speakers via an audio link. This mode of operation has also been used by corporations for employee training and employee relations programs. A number of small group sessions at the different locations using taped material is followed by a live video teleconference featuring the senior officers of the corporation, again with audio feedback. This form of teleconferencing is available through INTELSAT's occasional use of television service.

A second form consists of two-way interactive video and audio which is designed to take the place of an actual face to face conference around the table. The awareness that this form of teleconference is a substitute for business travel is increasing rapidly in major corporations round the world, and the convenience of a teleconference is sometimes crucial where busy senior executives must exchange information and views. Because of the privacy requirements attendant to business meetings, these teleconferences usually require some form of encoding of the signals prior to transmission, and this encoding is easily accomplished using digital techniques. INTEL-SAT has already demonstrated this new field at a number of satellite conferences where a number of international sessions have been linked through videoconferencing around the world.

International Business Service (IBS) is a digital service introduced in 1983 to meet the needs of the business community and designed to carry all types of telecommunications services, including video teleconferencing, high- and low-speed data, facsimile, packet switching, non-public-switched telephony, and electronic mail for business premises. IBS provides a speed range from 64 kbps to 8.448 Mbps with features attractive to the business user, including customer premise earth stations.

Following its introduction in the North Atlantic in 1983 on existing 14/11 GHz capacity, the service is being expanded to other regions to provide a complete global coverage. In addition, the last three INTELSAT V satellites were modified to provide a more effective service in the 14/12 GHz frequency band.

The main features of IBS include

Totally digital service

High quality and global in coverage

Flexibility

Small or medium size antennas

International and/or domestic

A modern service to meet contemporary needs

INTELNET, a service for data broadcast and collection from small earth solutions was introduced in 1984. This uses small inexpensive terminals working to or from a large central hub earth station. Applications range from networks for sending out financial news to networks collecting data on oil and gas exploration rigs, gathering environmental data, and inventory control networks. INTELNET I networks are one-way data distributed or broadcast from a large earth station to a number of small microterminals.

Typically, the large earth station is a Standard A (15 to 30 m) and the receiving terminal can be as small as 0.6 to 0.8 m with a down-converter and control module.

INTELNET II networks can be used for data collection or two-way purposes. These services are provided on a full or fractional transponder lease basis.

A thin-route telephone service, known as VISTA was introduced in 1982 to provide a basic service to rural or remote communities that have inadequate or no telecommunications services. Unfortunately, even with the basic satellite service available, and however low the tariff, the cost of the small basic earth station is often too high for many of the situations where it is most needed, and the response to this service has been slow to materialize.

The service can be used either domestically or internationally in, or among, countries with a relatively small requirement, or by special development projects where there is a need for basic telecommunication to a remote site.

Two new earth station standards were developed especially for this service, the Standard D-1 and D-2. The Standard D-1 station has less stringent characteristics than other standard INTELSAT stations, is about 4.5 to 5 m in diameter, and has the minimum of associated electronic equipment. These earth stations work usually to a Standard D-2 or a Standard B earth station which are both about 11 m in diameter.

This service is provided on a per channel basis rate either at nonpreemptive or preemptive rates. The first VISTA circuits were brought into service in July 1985 as a thin-route private business application between the United States and Madagascar (see Chapter 5). A few months later Australia started to use the service to some small and remote islands in the Indian ocean (see Chapter 9) VISTA has been slow to develop mainly due to the cost of the earth station in areas where revenues would be very small.

Ever since the earliest days of INTELSAT, satellites have been available for use when major terrestrial cable or microwave systems break down, and special arrangements exist whereby preplanned implementation is possible at short notice. The cable circuit restoration tariff has a daily rate.

As the number and capacity of these terrestrial systems has grown, satellite backup has also had to expand. There are now facilities for restoring medium and large capacity cable systems using full satellite transponders for this purpose. It is no longer economic, however, to maintain transponders specifically for this eventuality and these services will be linked with other services that can be preempted if the need arises.

2.2. INTERSPUTNIK

2.2.1. The Organization

INTERSPUTNIK was first established in 1971 as an international satellite organization which can be joined by any sovereign state, not just members of the ITU as in the case of INTELSAT. It is, however, inherently very much linked to the communist part of the world with its headquarters in Moscow. INTERSPUTNIK started service in 1974 and now has 16 members.

The INTERSPUTNIK Board is the main governing body and is directly responsible for the establishment and development of the INTERSPUTNIK system. Each member country is represented on the board and is allowed one vote.

The board determines the technical needs for INTERSPUTNIK communications satellites and earth stations, authorizes the assignment of satellite capacity, and appoints the Auditing Committee, which supervises the financial and economic aspects of the organization. A Director General, presently Mr. Boris Chirkov, is selected by the board to preside over the INTERSPUTNIK Directorate, a permanent executive and an administrative staff of only 40 which implements the decisions of the board. The board conducts meetings at least once a year held, in turn, within each member country. The U.S.S.R. is the prime policy maker at these meetings that are understood to be much less participative than the meetings of INTELSAT.

NonSoviet bloc users include Algeria, Iraq, Israel, Libya, Syria, and the United States. The United States' first large use of INTERSPUTNIK was the Turner Broadcasting reception of the Goodwill Games in 1985, but the signs are that reception of Soviet television will increase in the United States. Members of INTELSAT that use INTERSPUTNIK are required under the INTELSAT Agreements, to report and coordinate their use of INTERSPUTNIK under Article XIV(d) to ensure technical compatability and assessment of economic harm to INTELSAT.

The INTERSPUTNIK organization leases transponder capacity on three of the U.S.S.R.'s Statsionar satellites for transmitting television, telephony, and telegraphy. These satellites are located over the Atlantic ocean at 346°E, and over the Indian ocean at 53° and 80°E. News and television programs include selected programs taken off the INTELSAT satellites.

INTERSPUTNIK member nations with geographic locations that cannot access the same Statsionar satellite can intercommunicate through an earth station within the central zone common to both orbital satellites. This earth station then relays programs from one satellite to the other, or uses terrestrial facilities to pass traffic from one country to another.

INTERSPUTNIK telephony and telegraph traffic is transmitted using SCPC modulation techniques. Like INTELSAT, each member, or user country, owns and operates its own earth stations, which typically are about 12 m in diameter and have a figure-or-merit (G/T) of 31 dB/K.

2.2.2. INTERSPUTNIK Satellites

Like INTELSAT, INTERSPUTNIK uses several different classes of satellites each with its own distinct capabilities. The geostationary Statsionar spacecraft series are called Raduga (meaning rainbow), Gorizont (meaning horizon), and Elkran (meaning screen) satellites. Although the first Statsionar satellite was not launched until 1975, the U.S.S.R. launched 41 in the 10 years to 1985. The U.S.S.R. also uses some nongeosynchronous satellites called Molniya (meaning lightning) to reach some remote places and military installations north of the Arctic circle that have limited or no access to geostationary satellites.

Gorizont satellites have considerably higher down-link power than most western satellites, presumably to improve service for the news and television broadcasting requirements which can be accomplished due to the lesser requirement for voice and commercial traffic.

Most western 6/4 GHz satellites transmit less than 10 W of power per transponder, whereas even the older Gorizont satellites have 15 to 40 W TWTAs and global, hemispherical, zone, and spot beam capabilities with each coverage pattern selectable by control from the ground. Power levels range from a minimum of 26 dBW at the coverage edge of the global beam to a maximum of 48 dBW at the center of the steerable spot beam. This higher power satellite could prove very attractive to lesser developed countries that have been unable to persuade INTELSAT management to adapt a similar approach.

Like many American-made satellites, they use 36 MHz transponders in the C-band frequency range. Their beams are contoured for global and northern hemispheric coverages and use circular polarization with somewhat higher down-link power than the INTELSAT satellites. On account of their higher power, they only have six tranponders each of 34 MHz bandwidth and much less capacity than the Western equivalents.

A new series of satellites called TOR is expected to be introduced in 1990 and eight are planned to provide almost full global coverage. It is reasonable to suppose that these will also greatly increase the global scope of INTERSPUTNIK. The TOR satellites will use a new 42.5 to 45.5 GHz up-link and 18.2 to 21.2 MHz down-link for fixed land services, and also mobile services for land, aeronautical, and maritime uses. These satellites will be able to offer a wider range of services than those of INTELSAT and INMARSAT combined—at least for the world's communist affiliates. This may attract more countries to the INTERSPUTNIK system than there are at present.

The coverage of the TOR satellites from the eight orbital locations (8, 23, 35, 45, 85, 128, 190, and 333.5°E) will provide a top quality global broadcast system.

2.2.3.   INTERSPUTNIK Earth Stations

Again, like the INTELSAT system, each country owns and is responsible for its own earth stations, but the INTERSPUTNIK earth stations are smaller due to the lower capacity and higher down-link power of the satellite.

An INTERSPUTNIK earth station carries much less traffic than the typical INTELSAT earth station, and is very similar to an INTELSAT Standard B earth station having an antenna diameter of about 12 m and a G/T of about 31 dB/K.

The two main earth stations for INTERSPUTNIK are located at Vladimier and Dubna. Dubna is also the main INTELSAT earth station for the U.S.S.R. and is just north of Moscow; it was built originally for the 1980 Olympic Games. The Vladimir earth station is purely for INTERSPUTNIK and is situated over 200 km east of Moscow but information and full details of the INTERSPUTNIK system are not readily available.

2.2.4. INTERSPUTNIK Services

For telephone, single channel per carrier (SCPC) modulation techniques are used through a single satellite transponder, which is a very small traffic volume compared to the voice traffic used in the INTELSAT system. Some TDMA equipment is now being introduced for larger (12 to 60 channel) requirements. Just over 20 countries are believed to use the system's telephony services.

For television, two television channels are now transmitted on a daily basis from Moscow through four GORIZONT satellites to locations in the U.S.S.R. (includes 11 time zones) and to other INTERSPUTNIK users who wish to receive them. The transmissions originate from the U.S.S.R.'s Ministry of Communications as Programmes I and II, which are mainly received by 3 m antennas for local cable or rebroadcast distribution networks. When the TOR satellites are introduced around 1990, INTERSPUTNIK television broadcast capabilities are, however, likely to surpass those of any Western satellite system.

INTERSPUTNIK's present charges are not known, but the Organization claims they are much lower than those of INTELSAT, and the earth station cost is also claimed to very much lower due to the higher power satellite down-link beams.

2.3. INMARSAT

2.3.1. The Organization

The birth and growth of the INTELSAT network stimulated interest among maritime nations as communication between, and with, ships at sea had always been difficult and often impossible. In 1966, the Inter-Governmental Maritime Consultative Organization (IMCO) began studying the operational requirements for a maritime satellite system, recognizing its potential improvements for safety of life at sea through improved communications. IMCO, in 1968, also started to consider using the same satellites for aeronautical purposes.

The IMCO Assembly of 1973 decided to convene an international conference on the establishment of an International Maritime Satellite Organization (INMARSAT) and, in 1976, the INMARSAT Agreements were adopted by this conference but it had no existing or planned satellites with which to start operations. However, suitable satellites were available for lease from the United States' MARISAT system, the European Space Administration, and modified INTELSAT satellites.

While all this was happening, the United States was endeavoring to create a commercial organization to institute maritime satellite communications. About the same time in February 1976, the MARISAT system of satellite communications had been inaugurated by a U.S. joint venture consortium run by COMSAT General in conjunction with RCA, Western Union International, and International Telephone and Telegraph (ITT) with the launch of a MARISAT satellite over the Atlantic ocean. It entered service commercially in the same year, followed by another over the Pacific ocean, and a third over the Indian ocean. By the end of 1979, the first global system of commercial mobile maritime satellite communications was operational by the United States. These three satellites were still operational 10 years later in 1986.

Like INTELSAT, INMARSAT is based upon two agreements. The INMARSAT Agreements are termed the Convention and the Operating Agreement. INMARSAT's purpose is to provide space segment maritime communications, thereby assisting in improving maritime communications and also improving distress and safety of life at sea communications, efficiency and management of ships, maritime public correspondence services, and radio determination capabilities. The organization seeks to serve all areas where there is need for maritime communications and acts exclusively for peaceful purposes.

INMARSAT satellites provide telephone, telex, data, and facsimile, as well as distress and safety communications services to the shipping and offshore industries. Users of the system include oil tankers, liquid natural gas carriers, offshore drilling rigs, seismic survey ships, fishing boats, cargo and container vessels, passenger and cruise liners, ice-breakers, tugs, and cableships.

Although INMARSAT is known mainly for its maritime communications, it has ambitious plans for moving ahead into the wide open market place for mobile communications, including aeronautical and land-based mobile markets that have never been seriously considered by INTELSAT. INMARSAT comprises three bodies: The Assembly, the Council, and the Directorate.

The assembly is composed of representatives of member countries (parties) and each one has one vote. It meets every two years. The functions of the assembly include the consideration and review of activities, purposes, general policy and long-term objectives of INMARSAT, and the expression of views and recommendations thereon to the council.

The council functions in a similar way to the board of a company and consists of representatives of the 18 signatories with the largest investment shares, and four others elected by the assembly on the principle of a just geographical representation, and with due regard for the interests of developing countries. It meets at least three times a year and each member has a voting power equal to its investment share. The council has responsibility for providing the space segment necessary for carrying out the purposes of INMARSAT in the most economic, effective, and efficient manner, and also oversees the work of the directorate. The directorate has its headquarters in London, and houses the organization's permanent staff. The staff comprises under 200 people from over 30 countries and is headed by a Director General.

The organization is financed by the signatories each of which has an investment share based on its actual usage of the system. There are now 72 countries having over 4000 ship terminals registered under their flags.

For ships at sea INMARSAT has encouraged development of several ship terminals that are type-approved for installation on board ships and made by a wide variety of international manufacturers.

INMARSAT's ability to provide satellite communications to ship-borne earth stations has naturally led it into the field of supplying services to very small earth stations on land that is traditionally the bailiwick of INTELSAT. INMARSAT's facilities have been used on a number of occasions for fast disaster relief where a very small earth station can be lifted into a stricken area to provide emergency communication. INMARSAT facilities have also been used in such applications as off-shore and on-shore drilling rigs, and similar situations where an earth station is needed on a short-term basis. The potential market for these applications has been recognized by some manufacturers who now offer transportable systems that can be carried to any given location in two suitcases. This type of earth station is extremely attractive to many businesses, both large and small, as they permit direct communications to and from places with which it was hitherto impossible to communicate. Internationally, however, there are serious implications as such communications are often likely to be illegal. Most countries do not permit use of such earth stations as they can adversely affect national telecommunications revenues. They also circumvent many national laws and administrations and could also be used against the country's national interest which is why many countries are not eager to allow use of small private earth stations.

2.3.2. INMARSAT Satellites

INMARSAT started operations by leasing satellite facilities from INTELSAT, the European Space Agency (ESA), and COMSAT General of the United States. Three different space segment packages were used to provide

TABLE 2.2 **Orbital Status of INMARSAT in 1988**

	Atlantic	Indian	Pacific
Operational	MARECS, B-2	INTELSAT V, F5	INTELSAT V, F8
Launched	November 1984	September 1982	March 1984
Location	334°E	63°E	180°E
Spare	INTELSAT V, F6	INTELSAT V, F7	MARECS A
Launched	May 1983	Octoer 1983	December 1981
Location	342.5°E	66°E	178°E
Spare	MARISAT F1	MARISAT F2	MARISAT F3
Launched	February 1976	October 1976	June 1986
Location	345°E	73°E	176°E

the initial worldwide coverage, but INMARSAT is now procuring its own satellites for the second generation due in 1988.

Near worldwide coverage is obtained by providing a three satellite system with satellites located centrally over the Atlantic, Indian, and Pacific oceans. Present deployments are shown in Table 2.2.

The MARISAT satellites have well exceeded their expected life and INMARSAT signed a three year lease extension for the three MARISAT satellites in October 1986 for US$5.4 million as backup for existing satellites. The system will continue to operate using INTELSAT MCS and ESA's MARECS satellites until the INMARSAT satellites become operational in 1989.

The Maritime Communications Systems (MCS) are special maritime packages built into five of INTELSAT V satellites. These units are leased from INTELSAT for US$4.675 million per year.

The two MARECS satellites are leased from ESA and are a development of an earlier MAROTS experimental spacecraft.

The second generation spacecraft is being supplied by an international manufacturing consortium headed by British Aerospace. As well as expanded maritime communications capacity, these satellites are also equipped to provide communications to aircraft. These are due to be operational late in 1989 and a third generation, now at the procurement stage, is planned for 1994.

2.3.3. INMARSAT Earth Stations

There are two distinct types of earth stations in the INMARSAT system, the fixed earth stations on land, called coast earth stations (CES), and the small mobile earth stations used on board ships (SES).

The coast earth stations are owned and operated by each individual country where they are located and are usually owned by the national telecommunications administration. Calls from land are connected through these earth stations to the ships at sea where they are received by the ship earth station. Stations are located in many different countries and there are several serving each ocean region giving a number of alternative paths for communications with ships. Coast earth stations usually have antennas of between 10 and 13 m diameter. Ship earth stations are designed for installation and operation on board ships and must work in extreme environments. The antenna is usually less than 1 m in diameter and is usually located high on a mast or the ship's superstructure inside a protective radome. The equipment must be capable of operating under extreme weather conditions such as hurricanes, high seas, and extremely low temperatures. Special gyro mechanisms keep the small antenna tracking the satellite regardless of pitching, rolling, vibration, and changes in the ship's course.

Other equipment, such as telephone, telex, and electronic equipment is housed below deck, usually in what is known as the radio room. Some ships have more specialized equipment for data transmissions and facsimile, or even television.

Each ship earth station has its own unique number, or call sign, for calling and dialing purposes. Each ship earth station therefore consists of two major parts; the above deck equipment (ADE) and the below deck equipment (BDE).

The above deck equipment, at present, typically consists of a parabolic antenna mounted on a stabilized platform, a diplexer to separate the transmit and receive signals, a solid state high power amplifier (HPA), a low-noise receiving amplifier (LNA), and a power supply. All the equipment is housed in a strong fiberglass radome for protection against weather and sea water.

The diameter of the antenna depends on the type of standard being used on the particular ship. The Standard A SES ranges between 85 cm and 120 cm, but designs are becoming smaller. The smaller antennas compensate for the lower antenna gain by using a higher power transmitter to derive the effective isotropic radiated power (EIRP) which is about 36 dBW. The antenna mounting pedestal consists of a stabilizing system, usually four-axis, that had sensors and a servo system. The stabilizing system keeps the antenna continuously pointing at the satellite. In a four-axis system, the two lower axes counteract for the pitch and roll of the ship; the two upper axes control the elevation and azimuth axis through a drive unit to counteract changes in the ship's course. Information from the ship's gyrocompass is used to compensate for the turning and yawing motions of the ship, and a combination of built-in static rate/level sensors and servo loops counteract the pitch and roll. An additional step-track system compensates for slow tracking while the ship is in motion. This unit consists of a tracking control system that monitors the received signal strength from the satellite, which serves as a reference for automatic corrections of azimuth and elevation angles toward the satellite. Although the tracking is usually fully automatic, manual tracking is possible from below deck if necessary.

The below deck equipment is housed either in a rack or desk-type console that holds the antenna control and electronic equipment. There are also remote controls located in other parts of the ship for emergency situations.

Distress capabilities play an important part in the design and operation of all ship-board equipment, so that when a distress button is pressed, or a special command is given by the telex keyboard, an automatic SOS message is transmitted to a Rescue Coordination Center.

There are numerous manufacturers of INMARSAT approved ship earth stations all of which meet the vigorous design criteria issued by INMARSAT in its Technical Requirements Document (TRD).

Each new SES model has to be type-approved by INMARSAT to check that it meets all these specifications. Each individual SES must also be commissioned and approved by INMARSAT which results from certain information being supplied to INMARSAT followed by some simple tests after installation. Earth station research and development towards very small ship/boat terminals continues to bring satellite communications available to even the smallest boats and even weekend sailors.

2.3.4. Summary of Services

INMARSAT maritime services include the traditional services previously handled over the high frequency (HF) radio waves such as:

Telegraphy
Telephony
Some telex
Distress (SOS)

Not only are the new services infinitely more reliable than those provided by HF radio, but also many new features have become standard. Telephone services now include automatic calling from the ship to the shore, and in some cases semiautomatic calling from the shore to the ship. The semiautomatic facility is often limited by the inability of the shore network, but many countries are improving their networks to overcome this.

Telex operations are now fully automatic in both directions worldwide. Data transmissions are now possible up to 2.4 kbps using the telephone channels, and higher speed data transmissions up to 56 kbps are now possible, including facsimile. SOS alerts are virtually instantaneous via a Rescue Coordination Center using priority connections in the satellite.

INMARSAT went under contract in 1986 to flight test three different, but compatible, sets of air and ground equipment in mid-1987. INMARSAT's aeronautical services, which are now in the final stage of development and test, will shortly provide low speed data for aircraft and airline operational functions such as meteorological, air traffic control, position, performance monitoring, and safety communications. Higher speed data and voice communications for passengers will also become available.

In addition to these services, many new services are being planned and studied including:

Automatic selective interrogation or polling
Smaller ship earth stations for smaller ships
Higher speed data transmission rates
Television
Computer services
Radio determination
Improved safety at sea communications
Expanded aeronautical services for aircraft

Some of the ways that the INMARSAT services are now being used include:

Estimated time of arrival
Navigation data, weather, and daily reports
Cargo stowage planning
Reports, maintenance and spare parts
Private calls
Seismic and oil drilling data
Slow-scan television
Fleet management and crew rostering
Ship and travel agency coordination
Medical advice
News reports and teletext

2.4. EUTELSAT

2.4.1. The Organization

The European Telecommunications Satellite Organization, better known as EUTELSAT, was established in 1977 at the initiative of CEPT. Although final ratifications delayed the final establishment of EUTELSAT until 1985, a General Secretariat had been established in Paris in 1978 and had prepared all the groundwork for technical, financial, operational, and administrative matters.

After INTELSAT and INMARSAT, EUTELSAT is the world's largest international satellite telecommunications system. The organization is, in many ways, modeled on INTELSAT in that EUTALSAT's structure and the way it operates are governed by two international agreements, the Convention establishing the organization, and the Operating Agreement which lays down rules for the technical management of the satellite system. The signatories to the operating agreement are either the governments themselves or the public or private telecommunications entities given mandate by those governments, depending on the legislation of each particular country.

EUTELSAT owns, or leases, the space segment that is uses and is financed by the contributions of its members and by its operating revenues. Each signatory to the operating agreement makes a contribution to capital requirements in proportion to its investment share which is based on the use that signatory makes of the EUTELSAT space segment. Shares are adjusted each year, but for the first four years (at least) after the first satellite was brought into service (1985) investment shares are fixed by the board.

Users of the EUTELSAT space segment pay utilization charges, which are designed to meet the organization's expenditure on its administration, operation and maintenance, establish a working capital, amortization, and provide a return on the capital invested by the signatories.

The organization's main purpose is the design, development, construction, establishment, operation and maintenance of the space segment of the European telecommunications satellite system. The EUTELSAT space segment is intended primarily for the provision of international public telecommunications services in Europe. However, the convention does stipulate that should the need arise, the EUTELSAT space segment can be used for domestic telecommunications services of the member countries, or for specialized telecommunications services, international or domestic.

As its name implies, EUTELSAT is a European organization with European membership only. The basic principles of EUTELSAT are very similar to those of INTELSAT:

1. The objective of the organization's members is to pursue the establishment of telecommunications satellite systems as part of an enhanced European network. Membership of EUTELSAT is open to the CEPT countries and to any other European state belonging to the International Telecommunications Union (ITU).

2. All members of EUTELSAT have equal access to the space segment. European countries that do not belong to the organization can also have access to the space segment.

3. The system established and operated by EUTELSAT is to enable its members to offer more extensive telecommunications services, with the aim of developing relations between the peoples and economies of the countries that participate. EUTELSAT also has to operate on a sound economic and financial basis, having regard to accepted commercial principles.

4. The EUTELSAT system complies with international agreements or conventions with regard to use of the radio frequency spectrum and

the geostationary satellite orbit. The ITU acts as coordinator in this area to ensure the most effective and equitable use of frequencies and orbital positions. EUTELSAT also takes into account the recommendations of the CEPT.

EUTELSAT has three principal organs:

1. An Assembly of Parties composed of representatives of the member states. The Assembly of Parties expresses its views or makes recommendations to the Board of Signatories, particularly regarding the organization's policy and long-term objectives. Meetings are normally held every two years.
2. A Board of Signatories comprising the representatives of the user administrations of the satellite system. The board's principal task is the design, procurement, establishment, operation, and maintenance of the space segment. The board prepares and verifies the budgets and accounts and fixes utilization charges for the space segment, prepares traffic forecasts, determines conditions for access to the space segment, approves earth stations, and is responsible for coordination between satellite telecommunications systems and for interconnections with the terrestrial networks. The board meets several times each year.
3. An Executive Organ headed by a Director General appointed by the Board of Signatories. The Director General's term of office is six years, and this can be renewed. The Director General is the chief executive and legal representative of EUTELSAT. He is directly responsible to the Board of Signatories for the performance of all functions of the executive organ and for the structure and employment of its staff which is presently about 200. The first Director General is Mr. Andrea Caruso of Italy who had previously been a Deputy Director General of INTELSAT.

2.4.2. EUTELSAT Satellites

EUTELSAT I Series. EUTELSAT procures its satellites from European sources and the first series of EUTELSAT spacecraft were developed from the European Community Satellite (ECS) and the Orbital Test Satellite (OTS) which were Europe's first attempts at operational satellites. Five satellites were procured for the EUTELSAT I series of which four have so far been launched successfully. Each satellite in this present series has 19 transponders of which 13 are used for television although several transponders have failed and there are no spare TWTAs remaining in either satellite. The typical European coverage provided by EUTELSAT is shown in Figure 2.13.

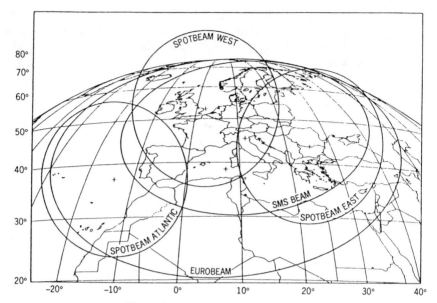

Figure 2.13. EUTELSAT coverage: Eutelsat I.

EUTELSAT II Series. EUTELSAT expects to need 30 transponder satellites to meet its forecast demands during the 1990s, but such a satellite would need too much development work in the time available so the EUTELSAT II series is an intermediate generation of five satellites planned to enter service in late 1990 to replace the expiring EUTELSAT I satellites. There are options for two additional satellites.

EUTELSAT II satellites will operate in the same Ku-band as the first series but will have only three down beams which will cover the European Broadcast Union (EBU) service area, a West European beam, and an Atlantic beam to provide coverage to the Portuguese, Spanish, and Nordic Atlantic islands.

As a result of the increasing demands for television services, the design of EUTELSAT II was changed in September 1986 to provide wider reception by smaller TVRO antennas, and increase the number of television channels to 16 with plans to lease eight of them for television.

The change means that EUTELSAT will be able to provide transponder services similar to Luxembourg's Astra satellite, and others, which plan to have similar 16 channel satellites ready in 1988. The modified EUTELSAT will then be the most powerful beam over most of Western Europe although national direct broadcast satellites (DBS) may provide higher power to individual countries within Europe.

2.4.3. EUTELSAT Earth Stations

The ground segment consists of a small number (14) of trunk earth stations with 15 to 20 m antennas, somewhat more (60) smaller earth stations with 5 to 15 m antenna, and a large unknown number of TVRO earth stations. All the transmit and receive earth stations are operated by the member PTT Administrations. In addition, there are a growing number of small earth stations on customer premises for the Satellite Multi Services (SMS) that are very similar to the INTELSAT IBS earth stations.

2.4.4. Summary of EUTELSAT Services

Services include trunk telephony between the member PTT Administrations, television, international business services, and probably direct television broadcasting services in the early 1990s. Already EUTELSAT claims over 400 million users in countries between Ireland and Turkey.

International transit centers exist in each member country's public switched network and they are connected to an earth station in the EUTELSAT network so that part of the intra-European digital telephone traffic can be sent by satellite, if needed, as part of the trunk telephone network.

Television services have increased most of all, and the requirements of the EBU and the other TV broadcasting companies are met, to the extent possible, as they emerge. The EBU leases two transponders on an exclusive full-time basis which are used by EUROVISION network. EUTELSAT also leases transponders to its signatories for domestic TV programs, and in fact the EUTELSAT I F-1 satellite is used largely for that purpose. Transponder capacity is available to signatories for occasional transmissions of unscheduled events or for promotional experiments. Other TV users include Sweden (SVT 1 & 2), Norway (NRK), United Kingdom (Music Box), and a relay of the U.S. Worldnet Service.

A wide variety of business services are now available within Europe using EUTELSAT's SMS which is the equivalent and compatible with INTELSAT's IBS. This service is available on transponders 11 and 12. The service is also available on France's TELECOM I satellite under an agreement with the French PTT Administration.

With only three years of operational service, EUTELSAT's results have been much better than was originally anticipated and it now has an annual operating revenue of over US$50 million which should increase again very considerably as the new satellites become available.

2.5. ARABSAT

ARABSAT (The Arab Satellite Communications Organization) is a regional system covering the 22 member nations of the Arab League and is complementary to the facilities of the INTELSAT system used by the same countries.

At a meeting of the Arab Telecommunications Union (ATU) in 1974 it was decided to establish a dedicated regional satellite system as well as a separate organization for the new system's administration. Later, at a meeting of the Arab League Ministers in 1976, ARABSAT was created, with its headquarters in Riyadh, Saudi Arabia. It was the first regional system challenge to INTELSAT.

Like INTELSAT, ARABSAT is governed by an Agreement that governs its overall policy, and has its own General Assembly, Board of Directors, and an Executive Body headed by a Director General. Many difficulties were encountered by the organization in the nine years that were to follow before its first satellite was launched in 1985. Firstly, INTELSAT was not enthusiastic about the prospect of its first major regional challenge, and there were major conflicts between ARABSAT members themselves concerning the finance and management of the proposed system.

When the system was devised in the mid-1970s the region was riding on the crest of the oil boom whereas at the launch time in the mid-1980s, many of ARABSAT's members were at least taking a second look at such expensive projects. Even in the mid-1970s it was clear that such a project could only become a reality if the richer Gulf States provided most of the money. It was only after much wrangling that the investment shares were agreed.

ARABSAT employed the United States' COMSAT General Corporation as consultant for preparing the tenders for the three proposed satellites and two control stations. Two prime competitors emerged, the U.S. Hughes Aircraft Corporation and a European consortium MESH involving British, French, German, Italian, Dutch, and Spanish companies. The Hughes bid encountered problems when it found that Hughes was on the Arab boycott list for its Israeli connections and the prime contract was eventually awarded to France's Aerospatiale with European subcontractors and the U.S. Ford Aerospace and Communications Corporation (FACC). The total value of the contract was US$135 million excluding ground equipment, launch, and insurance costs.

Manufacturing problems soon arose due to political problems when the U.S. government stalled work being done by FACC on the grounds that the satellites could be used against U.S. national interests in the area. Other problems emerged when Egypt was suspended from the Arab League, and when the remaining members argued over the specifications, costs, and control of earth stations.

The first ARABSAT satellite was eventually launched in 1985 on board an Ariane rocket. The second satellite was launched a few months later on board a NASA shuttle and the third satellite is kept on the ground in case of emergency at a later time.

The ARABSAT satellites are medium size having a capacity of 8000 voice circuits, seven television channels, and one S-band channel for community television services. There are two control stations for ARABSAT with the primary station near Riyadh in Saudi Arabia and the second, backup, station near Tunis in Tunisia. Both have 13 m antennas for tracking, telemetry, monitoring, and control purposes with the same facilities, except for launch control.

Member countries were slow to provide earth stations to use the new satellite, and although the satellite was available early in 1985, it was not until the end of the year that the first two earth stations started traffic—less than 12 circuits between Jordan and Bahrain showing the small amount of regional telephony and trade traffic that the system is likely to carry. The trunk telephony earth stations all use earth stations similar to the INTELSAT Standard B but use slightly larger antennas of typically 13 m diameter. Much smaller antennas can be used for TV receive-only purposes.

From its earliest years ARABSAT encountered problems in regional politics and the continuance of these tend to minimize the commercial viability or political usefulness of the network. With the satellites having only a seven year design life, it is essential that service starts early and expands quickly after launch, and this has not been the case with ARABSAT—at least for the first satellite series which appears to be entirely unprofitable. Telephony services are small leaving most of the satellite capacity available for regional or domestic television lease services. Several member countries have leased transponders from INTELSAT for many years for domestic services and most of these countries are continuing to use INTELSAT, especially as the ARABSAT annual lease rate of US$800,000 provides little, if any, financial advantage.

The system is still finding it difficult to attract users, mainly because so few Arabs countries have actually installed earth stations. In late 1986 only a few of the 9000 available circuits were being used. Bahrain is only able to use 31 circuits to reach just six countries that had earth stations (Jordan, Oman, Tunisia, Saudi Arabia, Kuwait, and Mauritania). Broadcasting has been mainly limited to a rebroadcast of the 1986 World Cup Soccer. The only domestic satellite system using ARABSAT is Oman which transferred from INTELSAT late in 1986.

Apart from the earth station delays, Egypt, until 1989, was excluded from all ARABSAT activities and being potentially the system's largest user, this also impeded ARABSAT's development. At the present time at least, ARABSAT has been neither a political nor a financial success. Its lack of financial viability could well cause problems in launching the third (replacement) satellite when due in 1992.

2.6. ASETA—A SOUTH AMERICAN REGIONAL SYSTEM

A regional satellite system for the Andean countries of South America has been under consideration for many years by the Association of National Telecommunications Organizations for an Andean Satellite (ASETA). The

ASETA members are Bolivia, Colombia, Ecuador, Peru, and Venezuela. The satellite project has been called CONDOR, and a two satellite system is still planned, although they are still considering other options, including INTELSAT.

2.7. RASCOM—AN AFRICAN REGIONAL SATELLITE SYSTEM

Under other names such as AFROSAT, AFSAT, and PDIA-SAT, an African regional system has been considered for about 10 years. Africa is a continent that could well use is own satellite system, but economic and political factors have continually prevented it from becoming a reality—or even a firm plan. Some African countries were among the first to use INTELSAT for domestic satellite networks, yet some of them were ill-planned, and remain today as disconnected national units. Many conferences have been held with the objective of implementing an African Regional System; some have produced names such as AFROSAT (by the PanAfrican Telecommunication Union, PATU), AFSAT (by the PanAfrican Posts and Telecommunications Union), PDIA-SAT (by the European ECA), and now RASCOM, (Regional African Satellite Communications System), which is supported by 48 African countries. Another feasibility study overseen by the ITU is to be completed in 1990 financed by some European countries, the Organization for African Unity, the UNDP, UNESCO, and the ITU.

3

INTERNATIONAL SATELLITE
SYSTEM COMPETITORS

Since 1983, when U.S. deregulatory policy first suggested that competition in the field of telecommunications satellite services would be advantageous, several new companies emerged with the objective of competing commercially with the existing international cooperative organizations such as INTELSAT and INMARSAT, as well as with the U.S. sole supplier of such services, COMSAT. The U.S. administration decided that private competition would provide a wider variety of services at a lower cost to the end-users than were being provided through its signatory (COMSAT) and the international organizations. Whether this will prove to be the case remains to be seen, but the policy did speed up changes that were already beginning to emerge within the international organizations, and also provoked the larger signatories to those organizations to become more customer oriented. This change in U.S. government policy has greatly improved the services available to the end-users, although whether the newly emerged international satellites become commercially viable remains to be seen.

3.1. THE POTENTIAL FOR COMPETITION

INTELSAT's policy towards separate satellite systems goes back to the very formation of the organization when the Definitive Arrangements were being negotiated and the United States and European policies were so different. It is interesting to note that the policies of the United States and some other countries have been completely reversed in recent years compared to when the agreements were being negotiated some 25 years ago.

The Europeans negotiated for wording in the agreements that would permit the creation of regional systems, such as EUTELSAT; the United States did not want any system except INTELSAT over which it expected to have considerable, if not entire, control through being the organization's manager. The eventual outcome was a compromise that permitted separate regional or international systems subject to coordination and agreement with INTELSAT as defined in Article XIV(d) of the Agreement:

(d) To the extent that any Party or Signatory or person within the jurisdiction of a Party intends individually or jointly to establish, acquire or utilize space segment facilities separate from the INTELSAT space segment facilities to meet its international public telecommunications services requirements, such Party or Signatory, prior to the establishment, acquisition or utilization of such facilities, shall furnish all relevant information to and shall consult with the Assembly of Parties, through the Board of Governors to ensure technical compatibility of such facilities and their operation with the use of the radio frequency spectrum and orbital space by the existing or planned INTELSAT space segment and to avoid significant economic harm to the global system of INTELSAT. Upon such consultation, the Assembly of Parties, taking into account the advice of the Board of Governors, shall express, in the form of recommendations, its findings regarding the considerations set out in this paragraph, and further regarding the assurance that the provision or utilization of such facilities shall not predjudice the establishment of direct telecommunications links through the INTELSAT space segment among all the participants.

The European interpretation of this was that the Assembly of Parties would probably give positive recommendations so that separate international systems (such as EUTELSAT) could be established provided INTELSAT did not make a negative recommendation. The United States' interpretation was that any country or countries wishing to establish a separate system must justify their proposal in terms of all the obligations they undertook when they signed the agreements, and that these checks would prevent competitive international systems. In the first 20 years of INTELSAT, two regional systems were coordinated, and INTELSAT did make positive recommendations for the European EUTELSAT system and the ATU's ARABSAT system on the grounds that they were technically acceptable and did not do significant economic harm to INTELSAT. Another international organization, INMARSAT, was also created to handle maritime satellite communications, as well as numerous national satellite systems.

The present considerations for economic harm to INTELSAT include:

The likely effect on potential revenue losses
The likely effect on future utilization charges
The likely effect on future signatory investment
The likely effect on future INTELSAT costs
The current financial position
The future growth opportunities and options
Future planning

In addition to the international organizations that supply satellite services, there are now emerging private companies planning to provide satellite services on an international basis; indeed by the time this book is printed, the first such private satellite service is scheduled to be in service. The formation of these companies followed the deregulation of the United States' internal telecommunications and the fragmentation of AT&T. At the time of writing, the number of such companies is small and most of them are more conceptual than practical, often called "paper" satellites. These companies are described in this chapter together with their planned activities to the extent that they are known. They are described in the chronological order in which they appeared, but some companies that entered have already withdrawn their earlier applications to the Federal Communications Commission (FCC). One thing that should be borne in mind is that any international trader, in this case a communications carrier, must have someone with whom to trade, in this case a foreign government or its telecommunications agency. This can be extremely difficult when many other countries are highly regulated and do not wish private foreign companies involved with their telecommunications.

The reason for the arrival of so many private companies eager to compete with the international organizations, which are all on the whole efficient, financially fair, and without unreasonable profit motives, is in my own view, the same fast, easy profit that was made when fragmenting the AT&T monopoly. They see a profitable opportunity to supply much wanted services to large business customers, which most countries' telecommunications administrations were too slow to provide around 1980. Few, if any, of these new companies really have the background to be critical of INTELSAT, or to provide the same reliable services. Thus far, most remain purely "paper" companies and their real motivation come from the United States' recent deregulation policy and the fact that a private company was created by the U.S. Government, and given almost a complete monopoly of the country's international satellite communications as long ago as 1962. Deregulation of U.S. domestic telecommunications started in the early 1970s, and several new communications carriers emerged to compete with the few giants, thus sharing the profits of the large established common carrier services. The United States' open market policy towards satellite communications is a policy not accepted by many other countries, but pressures on other countries are now changing this position.

Very frequently over many years, many well-informed Americans have failed to understand that there is even a difference between COMSAT and INTELSAT. Consequently, INTELSAT has frequently been blamed for many of COMSAT's actions—or inactions—as far as the American consumer has been concerned. The emergence of many of these new satellite companies could well come from a desire to break up the monopoly of the United States' participation in INTELSAT rather than to break up the international cooperative itself.

A similar, but lesser, problem is emerging also on Europe where

EUTELSAT holds the reins on European satellites. Luxembourg, which has always played a leading role in private radio broadcasting to all Europe, also launched its own satellite in 1988 that can blanket Europe with privately sponsored television networks. Taking a slightly different approach to INTELSAT and its competitors, EUTELSAT broke off coordination discussions with Luxembourg's ASTRA satellite owner, Societe Europeanne des Satellites, although discussions resumed later and were successfully completed.

When one of these new small companies actually breaks through the legal and institutional barriers, it is likely that some of the big international and national terrestrial carriers such as AT&T, Hughes, Contel, and companies from other countries such as France Cables and the British Cable & Wireless may well also join the fray and set up their own satellite systems for international services, thus slowly breaking up the international cooperative venture. Most of the new companies are now searching for other countries with whom to communicate, or to lease or purchase transponders, and it will be interesting to see how many of them succeed. The economics of the satellite business have become chilled and it will be very hard for the smaller ones to survive. It is unlikely that many of these companies will actually launch satellites, although some may profitably sell their orbital licences to other, larger corporations. The large U.S. satellite operators are likely, however, to reap from the deregulation by being able to use their excess satellite capacity, or create more capacity, for international purposes which had hitherto been restricted to domestic uses only.

3.2. THE UNITED STATES' ORION SATELLITE CORPORATION

ORION was the first of a series of United States companies to threaten INTELSAT's international cooperative by offering to provide international satellite communications in competition to INTELSAT. ORION was incorporated in October 1982 by the State of Delaware with an authorized capital stock of 100,000 shares having no par value.

The corporation was formed by Thomas McKnight, a former attorney with the Federal Communications Commission (FCC) and the White House Office of Telecommunications Policy (OTP) under President Ford. When the political winds changed, he joined the Gannett Newspaper Group during which time he turned his attention to the need for more transatlantic services and obtained support from some well-known industry names such as Gustave Hauser (former Chairman of Warner Amex), Walter Morgan (former scientist with COMSAT), the Pittsburgh National Bank, the Centennial Fund, and Guinness Mahon Inc.

Orion filed an application for authority to construct, launch, and operate a communications satellite system with the FCC in March 1983. The system proposed was an international Ku-band satellite system with satellites located at 37.5° and 47°W, to be launched in 1986. Due to satellite launch and other problems this was later delayed until the early 1990s. One of the satellites will be a ground spare.

In September 1985, the FCC found the application to be consistent with the President's restriction and overall policy of 1984 and ruled that ORION was legally and technically qualified to proceed with system planning. The FCC also ruled that ORION met the minimal financial qualifications but asked for more details before permitting it to start construction of the satellites.

When the European governments started moving to a more liberal policy towards telecommunications and the advances made by a later competitor (PanAmsat), ORION started to reorganize its plans. In April 1988, ORION announced it had agreed with British Aerospace to collaborate in the construction, launch, and operation of a transatlantic satellite system. Prior to its launch, ORION will use INTELSAT and also existing cable facilities. In the United States, ORION has teamed up with the SoutherNet fiber optic cable system. An application was made to the FCC in mid-1988 to operate two satellites over the Atlantic ocean with a third as a ground spare at a total of 88 transponders and an estimated cost of about US$350 million including launches and insurance.

3.3. THE UNITED STATES' INTERNATIONAL SATELLITE INC. (ISI)

ISI became the second potential private supplier of international satellite services, and like ORION, requested similar authorities from the FCC in August 1983, to operate satellite services over the Atlantic using two orbital satellites and one ground spare.

ISI was created specifically for developing new satellite services for its principle shareholders, TRT Communications which held 43 percent of the ISI stock. TRT has since been acquired by Pacific Telecom Inc. (PTI) which also has major Pacific cable interests. Satellite Syndicate Systems Inc. held 15 percent of the stock and provides video common carrier services widely in North America and the Caribbean as well as to other broadcasting and cable TV operations. The third major stockholder was Kansas City Southern Industries Inc. (KCSI) holding 14 percent. KCSI is another conglomerate with interests in transportation, telecommunications, and data processing.

The application to the FCC in August 1983 was for two operating satellites to 304° and 302°E that would use Ku-band frequencies and use two separate beams for Europe and North America. The prime objective was to provide video distribution and data markets for customer premise earth stations. At least half of the system capacity would be sold outright and only 15 to 30 percent would be available on a tariffed common carrier basis.

The FCC ruled, in September 1985, that the ISI application was consistent with Presidential restrictions but that it could not provide message telephone services, as proposed, to interconnect with the public switched telephone services. Like ORION, the FCC concluded that ISI only met the minimum qualifications needed for the FCC to authorize start of construction, and gave the same conditional authority that it had granted ORION. During the last few years ISI has rested quietly, but is rising again now that the market has become brighter and others have set the precedent.

3.4. THE UNITED STATES' PANAMSAT CORPORATION

Another company incorporated in the State of Delaware in 1984, PanAmsat has 80 percent United States ownership and 20 percent Mexican. The original U.S. stockholders were partners in RFW Satellite Services of Greenwich, Connecticut, and were Rene Anselmo (60 percent), Frederick Landman (20 percent), and William Stiles (20 percent), all of whom were partners and officers of the Spanish International Network Inc. (SIN). The partnership in 1987 was changed to Rene Anselmo (95 percent) and his son-in-law Frederick Landman (5 percent). The 20 percent Mexican ownership was held by Emilio Azcarraga Milmo who was also President of TELEVISA, which operates commercial TV broadcast services in Mexico. He was also Chairman of the Board of SIN Inc. and on the boards of Galavision Inc., Pan American Television SA, and declared other investment interests in Mexico and California.

PanAmsat also claimed that the proposed system would continue U.S. leadership in satellite development and technology, promote new markets, provide new outlets for U.S. industry, and help the FCC in finding ways to accommodate the concerns of other countries. PanAmsat further claimed that the proposed services were not available from, or proposed by, INTELSAT.

In September 1985 the FCC found PanAmsat's application consistent with the President's guidelines both legally and technically, but they had now shown the full technical qualifications needed to be authorized to begin construction. The FCC gave the same conditional authorization that it gave to ORION so that PanAmsat could pursue further financing as well as operating agreements with other countries. Later in the same month, the FCC granted PanAmsat a limited construction waiver that allowed it to spend, at PanAmsat's own risk, up to US$10 million before 1st December 1985 to obtain an RCA 3000 series satellite in time to meet the launch reservation with Arianespace. This postponement was followed by others when the United States' shuttle launcher program came to a disastrous standstill in January 1986.

Also in January 1986, PanAmsat supplemented and amended its earlier applications by advising that it had reached a Memorandum of Understanding with the Cygnus Satellite Corporation involving satellite and orbital location changes. Cygnus was another paper satellite system having few assets other than its FCC license giving conditional authorization to operate a North Atlantic Ku-band satellite system, and was also owned by a few individual shareholders, James L. Peeler (22.7 percent), Jack A. Shaw (22.7 percent), Andrew Werth (22.7 percent), and Paul Singh of the new Overseas Telecommunications Inc. (31.7 percent). The arrangement with Cygnus basically provided that PanAmsat would merge some equity interests, acquire Cygnus' orbital slot at 45°W, and share some frequencies. This amendment was approved by the FCC in May 1986 for the purpose of providing regional services between the United States and Latin America using five C-band transponders. PanAmsat had proposed two satellites and one ground spare with satellites located at 315° and 303°E. It was intended to use 19 transponders for South American domestic services and to offer the remaining 5 Ku-band transponders for services between the United States and Europe. The FCC did not act upon the proposal to supply domestic services to South America, but permitted PanAmsat to launch the Ku-band transponders as "deadweight."

PanAmsat obtained its first foreign partner in May 1986 when Peru agreed to buy one transponder, which gave PanAmsat enough leverage to ask the United States to start technical coordination with INTELSAT. Coordination with INTELSAT proved to be difficult as INTELSAT had consistently fought against PanAmsat and other similar systems. Lack of substantive information on how the satellite would be used did not help as PanAmsat could only provide any information at all for five out of the 24 transponders. There were also problems in tying down who actually would be responsible to INTELSAT for the future behavior of the system, PanAmsat, COMSAT (as the U.S. signatory) or the Department of State (as the U.S. party).

While the coordination meetings with INTELSAT were taking place in November 1986, PanAmsat requested FCC permission to purchase Cygnus and therefore use both the Cygnus frequency allocations and its Ku-band capabilities. The request was approved early in 1987. The cost of the purchase was US$350,000 divided between the four owners of Cygnus. The coordination process with INTELSAT was completed and favorably approved by the INTELSAT Board of Governors in December 1986 subject to three conditions: The U.S. party (the Department of State) agreed that the FCC and COMSAT would require PanAmsat to observe the operational parameters and conditions of use of its network or space segment as set forth in a board document. Secondly, use of the PanAmsat space segment would not be permitted by any INTELSAT party unless that party had informed the U.S. party, via PanAmsat, that it has met its obligation under Article XIV(c) or (d) of the INTELSAT Agreement. Thirdly, the U.S. party agreed to take the necessary steps to correct any unacceptable interference into the designated INTELSAT transponders caused by PanAmsat transmissions.

In September 1987, the FCC gave PanAmsat permission to launch its satellite and PanAmsat continued to contract with Arianspace for a launch in the spring of 1988. At this stage, the prime owner (Rene Anselmo) was expensively commited to his plan to introduce a limited international system in competition with INTELSAT, but found it extremely difficult to acquire foreign partners with whom to communicate. Early in 1988 the FCC gave provisional approval for PanAmsat to provide Ku-band services to Europe. Shortly afterwards the British Government (INTELSAT's second largest member) direct its INTELSAT signatory (British Telecommunications International) to provide earth station facilities for other satellite operators, that is, PanAmsat. PanAmsat immediately announced a transponder TV lease tariff of US$1.2 million per year. PanAmsat is now actively pursuing other European partners and customers.

The estimated cost for the initial single satellite system is over US$150 million up to the time of launch. Now it is launched and in orbit there will be recurring costs for operating the system, and at the present time it appears impossible for PanAmsat to make a profit at its going annual rate of US$1.2 million per 36 MHz transponder. Failure of PanAmsat during its early years will doubtless eliminate many of the other entreprenuerial satellite investors, but will also have breached the international cooperative's monopoly.

3.5. THE UNITED STATES' FINANSAT (FINANCIAL SATELLITE CORPORATION)

FINANSAT filed its application with the FCC in May 1985 requesting authority to construct, launch, and operate satellites over the Pacific and Atlantic Oceans to provide two ocean interconnects and provide services not on a common carrier basis starting in 1988. FINANSAT requested orbital slots at 313° and 182°E, which would provide full coverage of the Far East, Australia, the Americas, and Europe from a central transfer point in the United States.

Each satellite would carry 36, 72-MHz C-band transponders, which would be for sale on long-term lease to selected customers such as large financial institutions. Early in 1986 the FCC gave FINANSAT the same conditional approval as it had given ORION. Like ISI, FINANSAT is now biding its time for an opportune moment to launch its satellites.

3.6. THE UNITED STATES' COLUMBIA COMMUNICATIONS CORPORATION (CCC)

CCC is a wholly owned subsidiary of the Columbia Astronautics Corporation and was formed in 1981 in Hawaii by some former astronauts. It was formed to develop new satellite communications markets and services. Its major shareholders and officers included Clifford Laughton (President), Donald Slayton, Michael Collins, and a Briton, David Baker.

Columbia first applied to the FCC for a domestic satellite system that was refused in 1985. Later in the year they filed another application to construct, launch, and operate a Northern hemisphere international satellite system over the Atlantic and Pacific oceans located at 311° and 195°E. The FCC did not accept this application based upon its freezing order referring to orbital locations in the Atlantic ocean region. Columbia later revised its application to cover only the Pacific region using a satellite at 165°E with one ground spare that would cover all the Pacific rim countries. Financing would be raised, wholly or in part, by the Bank of America.

The FCC, in December 1986, gave Columbia approval to provide Ku-band facilities on a noncommon carrier basis, not interconnected with public switched message networks. Columbia is required to provide the required second stage financial information before February 1989, and is still endeavoring to obtain foreign correspondents.

The company plans to use a satellite with 36 transponders of 36 MHz bandwidth and 8 transponders with 54 MHz bandwidth. There are still no firm launch plans. In its cost estimate, Columbia itemized satellite construction costs at US$110 million for two satellites (one ground spare), US$60 million for launch, US$10 million for preoperational engineering, US$18 million for a control earth station, and US$6 million for one year's operating costs.

3.7. PAPUA NEW GUINEA'S PACSTAR

Pacstar is a joint venture between the Post and Telecommunications Corporation of Papua New Guinea and Pacific Satellite Inc. which is a United States company in which TRT Communications holds the majority interest. TRT also owns interest in ISI (section 3.3) and in a proposed Pacific ocean fiber optic cable system. TRT has since been acquired by Pacific Telesis Inc.

PACSTAR originally proposed a two-satellite system to begin operations in 1989 using C-band for the Pacific islands to minimize rain attenuation and also to permit small 2.5 or 3-m earth stations. Ku-band would be used for trunk services between the United States and the Far East. This proposal was made jointly at the January 1986 Pacific Telecommunications Council Meeting by PACSTAR's representatives, Donald Jansky of Jansky Telecommunications Inc. and Danny Coyle of Papua New Guinea. The PACSTAR system would provide domestic and international services to the Pacific island nations and would be providing competition not only to INTELSAT but also to Australian and other satellite interests in the area. Later reports also indicated that an L-band package on the satellite would provide transpacific aeronautical services in competition to INMARSAT.

3.8. THE UNITED STATES' CELESTAR

McCaw Space Technologies Inc. (SpaceTech) filed an application with the FCC in 1985 to provide mobile and radio determination services to augment its numerous cellular, cable, and radio services. Later, in 1986, SpaceTech filed with the FCC again, but this time to construct, launch, and operate a two-satellite system named CELESTAR. This application was approved by the FCC in 1987.

CELESTAR I would be a Ku-band satellite located at 170°E over the Pacific for linking the United States, Guam, and Hawaii. CELESTAR II would be located at 70°E over the Indian ocean to link Guam with Asia, Europe, and Africa. There would be one ground spare, and each satellite would consist of 102 36-MHz active transponders and 17 spare transponders to be launched in 1991.

CELESTAR anticipate a system cost of US$380 million starting in 1992 and using two RCA-5000 satellites provided the necessary finance can be arranged and foreign correspondents found.

3.9. ASIASAT

In September 1986, Pan Am Commerical Services, a subsidiary of the Pan Am Corporation, announced plans to establish a Pacific regional satellite system using a satellite that it had just purchased from Western Union, the WESTAR 6 satellite that had been launched and recovered from space in 1984. The purchase price was US$20 million for the retrieved satellite from Lloyd's of London. In July 1987 the company also purchased Canada's Anik C-1 satellite, which had been an orbital spare, since launched in 1986, for about US$63 million.

Pan Am Commercial Services, which has given launch support to NASA at Cape Canaveral for many years, joined Johnson Geneva USA Ltd. who had bought the satellite, to form a new company named Pan Am Pacific Satellite Corporation. Johnson Geneva had been in business for three years designing low-cost networks for lesser developed countries, and in this new joint venture, hoped to sell satellite services to countries in the Pacific basin, many of which seem unable to afford any service that is not free, and many of which claim they cannot afford even INTELSAT's cheapest services. Pan Am Pacific continued to search for other venture partners and joined up with British Telecommunications International (BTI), Telesat Canada International, and Hutchinson Telecommunications Ltd. of Hong Kong who together proposed to provide domestic services to countries in the Pacific rim as well as the many islands. BTI and Telesat withdrew from the venture and a new consortium was formed with China's International Trust and Investment Corporation (CITAC), and Britain's Cable & Wireless Group. The new consortium's name was changed to ASIASAT, and the satellites are planned to be launched on China's Long March rocket during the first part of 1989. This new consortium will be able to provide additional services to the region and do so competitively with INTELSAT.

3.10.   **TONGASAT**

The Friendly Islands Satellite Company (Tongasat), through the Tonga government, applied to the IFRB for eight orbital allocations late in 1988. Tongasat is headed by a United States entrepreneur, Dr. Matt Nielson, a former director with INTELSAT who had previously headed a number of small future satellite companies in the United States. The company stated it intended to operate three satellites starting in the early 1990s. The eight orbital locations applied for are at:

105.5°E	Tongasat 1 from November 1991
115.5°E	Tongasat 1 from December 1992
121.5°E	Tongasat 3
131.0°E	No details were given
160.0°E	No details were given
164.0°E	No details were given
170.5°E	No details were given
187.5°E	No details were given

Under the present rules, these applications prevent other applicants from using these locations unless launches do not take place before 1994. The company later said that some locations could be used by other satellite operators on a leased or purchased basis. This action puts a slightly new technique into the politics of planning and organizing the use of the geosynchronous orbit.

A very small, private, entrepreneurial company, through the good offices of a very small island state, can claim space for eight satellites. Tonga, a nation of about 170 small islands, 270 square miles, and a population of some 98,000 people can never be expected to use this amount of space resource.

4

INTERNATIONAL SERVICES

This chapter discusses the broad range of satellite communications services that are available through the international satellite organizations mentioned earlier. More detail on the specific classes of service are described in later chapters.

4.1. TELEPHONE AND TELEGRAPH

Early satellite communications services, which are now often rather vaguely called conventional, were for basic telephone and telegraph based upon an analog 4 kHz high quality telephone channel. Each channel can be combined with others to provide a wider band channel for data or broadcasting purposes, or divided into a number of narrower band telex or telegraph channels.

In most countries, these services are supplied by a single telecommunications administration, partly or wholly owned by the country's government. With very few exceptions, these countries use INTELSAT for providing international satellite services of this nature, although there are now some regional organizations such as EUTELSAT, INTERSPUTNIK, and ARABSAT. INMARSAT is used for maritime services and is covered in Chapter 7. A few countries permit a number of competing telephone and telegraph companies to access INTELSAT directly, but this, up to the present, is the exception rather than the norm. These types of applications are shown in Figure 4.1 with the national organizations supplying telephone, telegraph, telex, and private leased circuit services to the end-users that may range from a private home telephone to a large company's communications center.

These traditional international services include:

Public switched telephone services
Telex
Alternate voice/data (AVD)
Leased private circuits

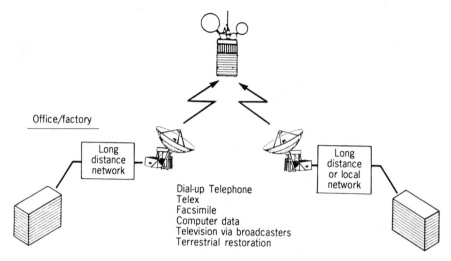

Office/factory

Long distance network

Long distance or local network

Dial-up Telephone
Telex
Facsimile
Computer data
Television via broadcasters
Terrestrial restoration

Figure 4.1. International satellite services.

These services were available in limited form prior to the arrival of satellites but are now so widely available that they are taken for granted as part and parcel of everyday life in almost every country of the world. The continuous and vigorous growth of this international tradiational traffic is shown in Figure 4.2.

Traffic routing is planned under the International Telecommunication Union's (ITU) Routing Plan, which has changed radically due to communications satellites and the large numbers of direct circuits that are now possible in the INTELSAT system. Now, in practice, most traffic routing is planned at INTELSAT's annual Global Traffic Meeting with some prior arrangements being made by large cable users to take these facilities into account. Global planners have always considered terrestrial and satellite systems to be complementary and have established alternate paths for diversity purposes. In a few cases, for some very small routes, it is still economically advantageous to route services via another point, which is termed "via routing." This is usually done through a mixture of terrestrial and satellite paths, but occasionally two satellite paths may be used, which is termed a "double hop."

In terms of satellite system performance and reliability, INTELSAT's practice is to record outages in the satellite link that have a duration equal to or greater than one minute. Although this length of outage is serious to many end-users, outages of less than one minute are rarely due to the space segment and are consequently not recorded for practical reasons. The satellite line link is defined as extending from the baseband test point of the radio equipment at the transmit earth station to the baseband test point of the radio equipment at the receive earth station. Degradation of service due

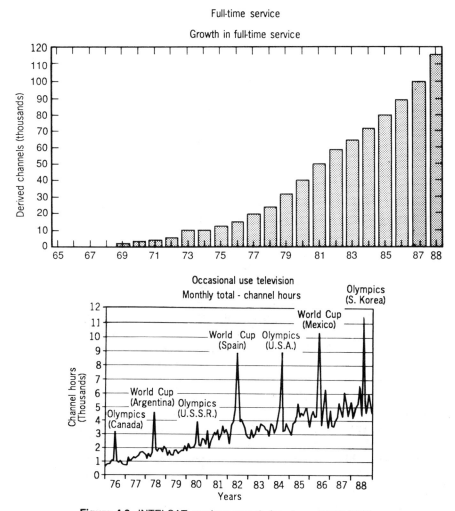

Figure 4.2. INTELSAT services growth (courtesy: INTELSAT).

to reasons such as propagation conditions is not recorded as an outage unless the service is degraded to such an extent that the circuit is considered unusable and taken out of service. Statistical information is maintained on availability and continuity of service, but this is becoming increasingly impractical due to countries' different standards and maintenance habits. The system was devised in the early days of satellite systems to measure the system performance; this has long since been proved and accepted as being considerably better than almost any terrestrial link, which usually consists of multiple links, each generating its own degradation hazards.

Availability performance is a measure of availability of an entire satellite or earth station. It is calculated from the duration of the system outages that interrupt all circuits through a satellite or an earth station. In the last 15 years, including all earth stations, this has never dropped below 99.75 percent.

Customer costs for these services vary widely although INTELSAT's charges to its signatories have dropped dramatically since services started in 1965 (Figure 2.1). The satellite charge forms normally only about 10 percent of the charge to the end customer with the remaining 90 percent being levied by the national carrier for the terrestrial link costs. With the present satellite channel charge of US$370 per month per voice channel, any future reductions by INTELSAT are likely to have zero impact on the charge to the end-user.

4.2. TELEVISION AND BROADCASTING

INTELSAT's services for high grade voice channels suitable for rebroad-casting have been available since the organization started and were, in fact, one of the first uses of the satellite for ceremonial openings. They are made up of ordinary 4 kHz channels, or multiples thereof, on an hourly basis through signatory earth stations. The service is available on a single or multidestination basis and a single destination voice broadcast channel can cost as little as US$65 for one hour from one country to another.

International television broadcasting was impossible prior to the advent of satellites except where countries were close together and had terrestrial links between them as existed in Europe. Like the conventional telephone and telegraph service applications, the distribution and reception of interna-tion news events by press, radio, and television media is now taken for granted in all but the least developed parts of the world. Once again, INTELSAT is almost the sole service provider although regional systems such as EUTELSAT and ARABSAT have recently emerged and some private competitors are in the wings.

Video and audio broadcast channels have been available on each INTEL-SAT satellite since 1965 for providing scheduled television services for news, sports, and other broadcast services. In the early days it was necessary to close down telephony services during these events, but the increase in demand for video services soon made it necessary for INTELSAT to dedicate specific transponders for these services, which became known as the "Occasional Use TV Service." This service is usually provided on each global transponder number 38 by means of two 20 MHz video channels at 6392/4167 and 6413/4188 MHz.

Television can be single or multidestination and double hop transmissions over two ocean regions are quite frequent. Television services are often complex due to the interface arrangements necessary for multiple destina-tion transmissions, particularly with video networks having different trans-mission standards. Orders for television programs are collected by the international carriers in various countries and passed on to INTELSAT's operations center in Washington D.C. where they are coordinated to ensure that satellite capacity is available at the time required.

Over time, the occasional use capacity became so congested, especially for special events, that extra capacity was needed and introduced to permit the occasional use service to continue providing quality service. The problem was first overcome by changing the transmission standard to permit two video carriers on the transponder rather than one.

Another means for keeping a useful and uncongested occasional use service was to provide attractive means for the larger TV users to stop using the service, and in 1981 an international leased television channel service was introduced allowing the lease of transponders for international television in units of 18 MHz. This service proved popular and was greatly expanded early in 1984 to cover a wider range of user needs. A wide variety of video services are now available for leasing with numerous priorities and tariffs which are discussed elsewhere, but their introduction successfully prevented saturation of the occasional use facilities.

INTELSAT has frequently been criticized for exorbitant television tariffs and in 1985 the Senior Assistant Secretary for Communications and Information wrote that "currently it costs a minimum of more than $2700 an hour to transmit television programming from New York to London using the facilities of AT&T, COMSAT, INTELSAT and British Telecom. Domestically, however, such service over a comparable distance relying on domestic satellites costs only about $790." (*International Journal of Satellite Communications*, Volume 3, 1985). The INTELSAT portion of this cost, assuming the most expensive occasional use service were used, was in fact only US$960 per hour showing clearly that the high costs lay with the terrestrial carriers rather than with INTELSAT. One hour of off-peak TV transmission between New York and London can cost as little as US$210 using INTELSAT. A survey performed by the European Broadcasting Union in 1986 showed very disparate charges by countries for passing on the INTELSAT service and some of these are given in Table 4.1.

At the end of 1987 there were 24 full-time leased television channels in service (786 MHz) and the occasional use services reached a record 55,778 hours during that year.

The occasional use service remains the most widely used means for televising live news coverage on a worldwide basis and some of the main features of the services on principal satellites at 325.5°, 335.5°, 341.5°, 60°, 63°, and 174°E are:

Minimum 10 minutes

Charged by the minute from US$16/minute for single destination TV

Charges increase for additional recipients

Single or multidestination transmissions

There are cancellation charges

There are early termination charges

TABLE 4.1   Global Charges for using INTELSAT Television (Extracted from EBU Statistics)

Country	First 10 Minutes US$	Additional Minutes US$
INTELSAT	80.00	8.00
Brunei	80.00	8.00
United States	180.00	15.89
Madagascar	203.16	20.36
Costa Rica	244.34	7.84
Canada	360.10	14.40
France	512.33	23.65
Iceland	669.97	23.65
Portugal	591.15	23.65
El Salvador	690.00	26.00
Colombia	650.00	20.00
Mexico	691.00	22.00
Peru	599.82	22.00
Venezuela	640.00	24.00
Spain	748.79	23.65
West Germany	748.79	23.65
United Kingdom	748.79	23.65
Greece	906.43	27.59
Italy	906.43	27.59
U.S.S.R.	906.43	27.59
Poland	945.85	31.53
Turkey	985.25	31.53
China	985.25	31.53
Saudi Arabia	1,000.00	31.00
Roumania	1,103.50	33.50
Pakistan	2,000.00	65.00

Services on lesser used satellites at 307°, 310°, 332.5°, 338.5°, 359°, 57°, 66°, and 180°E permit:

A lower tariff

Any number of recipients at no extra cost

No cancellation charges

No early termination charges

Peak and off-peak rates

For major events that are designated in advance such as Olympic Games, World Cup Soccer, the Hajj Pilgrimage, the U.S. and British Open Golf, and Wimbledon Tennis, off-peak rates do not apply, but this wider variety has greatly reduced the congestion in the original occasional use transponders to which many countries have sole access. There are many special events that are covered by the INTELSAT occasional use television service including:

World Cup Soccer
Olympic Games
Wimbledon Tennis
Regional sports events
Royal Weddings
Summit conferences
World Prayer days

The 1988 Olympics in Seoul, Korea. As it has done over the last 25 years, satellite communications provided the world with television and news coverage of these games. More than three quarters of the world's population are now able to watch the Olympic Games, and this has become possible almost entirely due to the INTELSAT system.

Seoul was designated the host city for the 1988 Olympic Games in 1981, and from that time on, INTELSAT and the Korean authorities planned together to make the telecommunications of the Games a worldwide success. Almost like a rehearsal for the Olympics, the Asian Games were held in Seoul during August 1986, which provided plenty of practice for the Olympic Games of September 1988.

Nine of INTELSAT's 13 satellites were used to transmit Olympic news from Seoul in a maze of networks which are shown in Figure 4.3. In Seoul, a special earth station was installed by the Korean Telecommunications Authority (KTA) in addition to its existing earth stations at Boeun and Kumsan, and INTELSAT staff provided on-site assistance at the International Broadcast Center in Seoul to help with booking arrangements and technical support. By the end of the Olympic Games, INTELSAT's satellites provided the rest of the world with over 7000 hours of occasional use television services to 64 countries.

Apart from the occasional use television services from Seoul, several broadcasting companies used INTELSAT's short-term leased television services to provide full-time coverage of the Games to their respective countries. There were eleven of these:

Korea to:	MHz leased	Beam	Satellite
Canada (CBC)	72	Ku spot	180°E
Canada for Mexico (Televisa)	72	Ku spot	180°E
Canada extended to Spain (TVE)	72	Ku spot	180°E
United States (NBC)	36	Zone	180°E
United States (NBC)	36	Ku spot	174°E
Australia (Channel 10)	18	Hemispheric	180°E
United Kingdom (BBC, EBU and others)	72	Zone	63°E
Germany (ARD and ORS)	72	Zone	66°E
Japan (NHK)	36	Ku spot	174°E
Japan (NHK)	18	Zone	63°E
France (EBU)	24	Zone	66°E

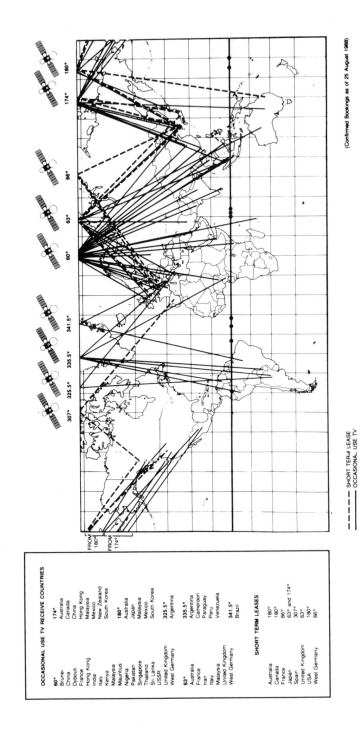

OCCASIONAL USE TV RECEIVE COUNTRIES

60°
Brunei
China
Djibouti
France
Hong Kong
India
Italy
Kenya
Malaysia
Mauritius
Nigeria
Pakistan
Singapore
Thailand
Sri Lanka
USSR
United Kingdom
West Germany

174°
Australia
Canada
China
Hong Kong
Malaysia
Mexico
New Zealand
South Korea

180°
Australia
Japan
Malaysia
Mexico
South Korea

325.5°
Argentina

63°
Argentina
Cameroon
France
Iran
Italy
Malaysia
Paraguay
Peru
Venezuela
United Kingdom
West Germany

335.5°

341.5°
Brazil

SHORT TERM LEASES

Australia   180°
Canada   180°
France   66°
Japan   63° and 174°
Spain   307°
United Kingdom   63°
USA   180°
West Germany   66°

FROM 180°
FROM 174°

307° 325.5° 335.5° 341.5° 60° 63° 66° 174° 180°

(Confirmed Bookings as of 25 August 1988)

– – – – – SHORT TERM LEASE
———— OCCASIONAL USE TV

Figure 4.3. 1988 Olympic Game satellite television coverage by INTELSAT (courtesy: INTELSAT).

Meeting of the U.S. and U.S.S.R. Heads of State in Iceland, 1986.
Iceland has only 150,000 telephones and 50 lines can normally easily handle
the daily traffic to the United States mainly through a single earth station, so
when Iceland hosted the weekend meeting between President Reagan and
Soviet leader Mikhail S. Gorbachev technicians had less than two weeks to
prepare for the communications onslaught. Every existing resource had to
be used and more resources had to be flown in, including transportable
earth stations and transmission equipment.

For White House purposes, special facilities exist for Presidential communications using COMSAT transportable earth stations and a permanent
earth station near the White House. The large rush of press reports and
television coverage were the main requirements.

During the first several hours, reportedly, the press had trouble trying to
phone home. Some calls were sent over AT&T's regular satellite circuits on
the international direct dialling network, but AT&T flew in a communications vehicle to establish 48 extra circuits using another transportable earth
station and a switching center in Alexandria, Virginia. All of the phone
costs, including AT&T's charge for flying in the equipment are routinely
divided among the news organizations who sent reporters with the President. ITT established extra facsimile circuits so that Japanese script could be
sent without the need for translation. IDB Communications of Los Angeles,
which normally transmits sports programs for cable channels, provided
television channels through a transportable earth station for six U.S.
networks.

Other U.S. TV broadcasters chartered cargo planes to carry small earth
stations to Iceland for transmitting their summit programs and transportation charges alone were around US$150,000.

INTELSAT used a record seven satellites to relay signals from the 10 or
more dishes that the networks flew to Iceland as shown in Figure 4.4. Other
countries that leased direct extra circuits from INTELSAT included France,
Iceland, Italy, Japan, Sweden, United Kingdom, West Germany, and the
U.S.S.R. Some statistics for these meetings are shown below.

	Ocean region satellites		
	Atlantic	*Indian*	*Pacific*
Hours transmitted	360	2.5	–
Number of transmissions	162	11	–
Participating countries	7	2	–
Earth stations transmitting from Iceland	**To**		
Skyggnir	United States, Andover		
Skyggnir	France		
CBS, McMichael	United States, Roaring Creek		
ABC, NewsHawk	United States, Roaring Creek		
CNN, Cresscomm 1	United States, Spring Creek		
STARS Terminal	United States, Andover		

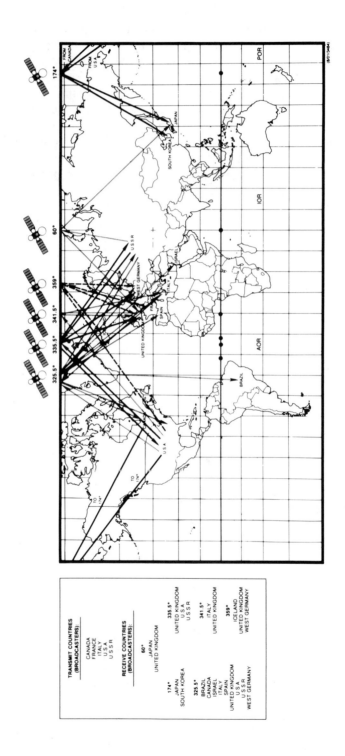

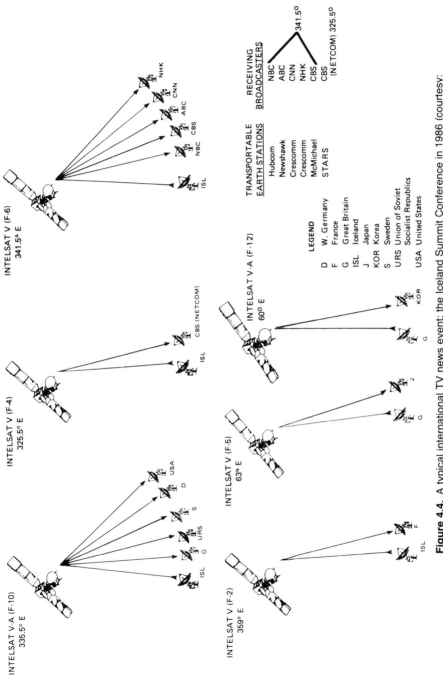

Figure 4.4. A typical international TV news event: the Iceland Summit Conference in 1986 (courtesy: INTELSAT).

93

World Cup Soccer 1982 and 1986. The World Cup is one of, if not the greatest, worldwide television event and seems more popular than even the Olympic Games (Figure 4.2). The most recent Cup took place in Mexico City between 31 May and 29 June 1986. Over 10,000 television channel hours were transmitted by INTELSAT alone and excludes those transmitted by other systems such as ARABSAT, EUROVISION, and domestic satellite systems such as the United States, Mexico, and Brazil. It is interesting to compare the INTELSAT World Cup television statistics for the years 1982 and 1986.

	1982	1986
Countries televising	92	75
Transmissions		
Atlantic	766	577
Indian	108	73
Pacific	8	86
Total	882	738
Minutes transmitted		
Atlantic	315,215	345,877
Indian	92,420	37,020
Pacific	1,975	60,496
Total	409,610	442,392

The reduced number of countries participating in 1986 was due to the ARABSAT countries mainly watching a retransmitted version through ARABSAT, some European countries viewed the EUROVISION rebroadcasts, and most of the island nations in the Caribbean watched the Cup on the United States' WESTAR domestic satellite system. In addition to the occasional use figures shown, there were also five full-time television channels leased between Mexico and the two largest recipients, Brazil and United Kingdom.

For events of this nature INTELSAT normally sends staff to the event location to assist in the smooth running of the satellite transmissions.

Live Aid and Sport Aid, 1985 and 1986. In 1985 and 1986 worldwide television was used for the first time by charity organizations to raise large funds for famine relief in Africa. The first such event was Live Aid in July 1985 when rock star celebrities gave concerts in major cities of the world, transmitted them by satellite to New York, and then the combined program was retransmitted to the world. These two worldwide television extravaganzas were arranged by VisNews International out of the United Kingdom and its U.S. partner Bright Star.

Sport Aid was produced by Global Media Lted. and was a joint effort by UNICEF and Bob Geldof, the Irish pop-star. Live Aid raised more the US$100 million and Sport Aid US$200 million for African famine relief.

Sport Aid was the largest satellite broadcast ever, and brought live television coverage of fund-raising running events in 13 countries on six continents. The two hour broadcast used 24 transponders on 14 satellites belonging to INTELSAT, EUTELSAT, RCA and Western Union in the United States, and Canada's Telesat. The combined network was an enormous undertaking and is shown in Figure 4.5.

Sport Aid naturally produced a lot of "firsts," including the first ever television to come out of Burkina Faso in West Africa, and the use of quad video to condense and route seven signals through London into two satellite paths. This produced a four-way split picture on two screens in New York which enabled the producers to select the required picture without delay. The organization of this event necessitated coordinating activities with INTELSAT, EBU, and companies or administrations in at least eight countries.

4.3. BUSINESS COMMUNICATIONS

Throughout telecommunications history, new services have followed the needs of the large businesses that require them to perform their trades cost-effectively. This has applied whether the trade be spices, or modern equivalents such as oil or high technology. Following this historical trend, it is the large businesses, mainly of North America and Europe, which have led the way for the international need for improved communications.

For several decades some businesses such as shipping, oil, and commodity traders have used leased circuits between their headquarters and overseas branches. These leased circuits were usually for direct two-way teletype with automatic error correction known as ARQ. Some of the larger companies such as SITA (International Society for Aeronautical Telecommunications) used worldwide networks of this type to meet their needs. Some major airlines and banks had similar networks arranged through a variety of administrations and private communications companies many of which used conventional INTELSAT telephone channels for the purpose.

International private business networks began to appear around 1960 by knitting together customer needs and fitting them into segments of the existing public network and leasing the networks on a full-time basis. These networks appeared before the advent of satellites, but the satellite link when it became available, greatly improved their quality and availability. In fact, almost the first commercial use made of INTELSAT was for a business network as we known it today—the global telecommunications network for the U.S. Apollo project to send man to the moon.

The Apollo project telecommunications network for NASA linked and provided voice, data, and video services between controls in the United

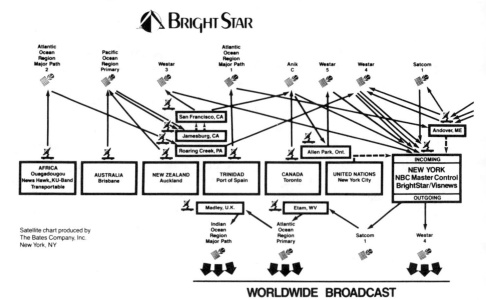

WORLDWIDE BROADCAST

Figure 4.5. The Sport Aid international TV broadcast in 1986 (courtesy: Brightstar Communications Ltd.). Land lines and connections are indicated by dotted lines.

States and tracking stations round the world including Carnarvon (Australia), Hawaii, Ascension Island (U.K.), Canary Islands (Spain), South Africa, and also ships at sea.

The next business systems were for airline operations, but other businesses quickly followed suit such as shipping, oil, banking, and trading commodities. These business groups felt, at the time, that they were engaged in common activities and could pool their requirements to obtain more favorable rates. Most countries resisted these user groups at the time although nearly all of them eventually yielded and found even more profit in the new private business networks. INTELSAT's conventional telephone channels played a major role in the development and growth of these early private business networks.

The requirements for international leased circuits have usually been hard to identify or quantify, and the first known attempt to do this on a wide scale was the United States' National Telecommunications Information Agency (NTIA) together with some of the country's international record carriers. The first report of this study in 1985 showed that the requirements for international private leased circuits were highly concentrated among a relatively small number of business users; a logical conclusion! These studies were, however, limited to the likely requirements for international leased circuits by United States businesses; foreign international lease requirements were not covered by the study which was of a national nature only. Later studies of business requirements in other countries were performed revealing a great need for extension of international business networks.

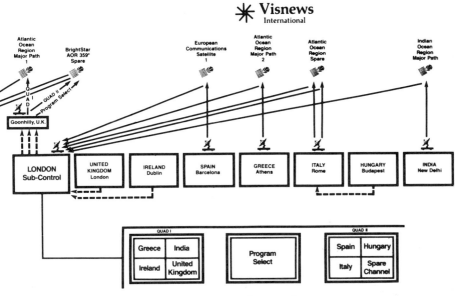

Figure 4.5. (*Continued*)

Satellites have made leased transponders available, not only to large international carrier companies, but also to any other corporations that feel the need. It is not therefore surprising that when spare satellite capacity has been available, large businesses have been guided towards the attractiveness of these facilities leading to much higher market needs than could be met by the conventional telephone channels.

Since 1981, companies across the United States and Canada have been turning to satellite communications, or to fiber optic cable facilities for general communications and long distance teleconferences. These facilities are being found to be very cost-effective in many cases and reduce costs of travel, market presentation, inventory control, and administrative costs. Although many of these applications are domestic in nature, many of the same users are also involved in international trade and are turning to satellites to meet those needs, which are discussed in more detail in Chapter 5.

4.4. TERRESTRIAL BACKUP SERVICES

Since INTELSAT's earliest days, satellite channels have been used, as needed, to back up failures in terrestrial networks whether they be cable or microwave systems. Early in the evolution of INTELSAT, special arrangements were set up whereby preplanned restoration is available at short notice. A daily cable circuit restoration tariff was introduced and whenever a terrestrial system fails, services can be transferred quickly to a satellite at a daily rate of US$28 per voice circuit. The quantity of cable traffic restored over the last 12 years is shown in Table 4.2, and the types of restorations that occurred in 1987 are shown in Table 4.3. The present generation analog submarine cable typically has a capacity of around 4000 circuits and the maximum number of circuits restored by satellite for any one cable has not

TABLE 4.2   Cable Restoration Statistics

Year	Circuit Days			
	Atlantic	Indian	Pacific	Total
1974	15,412	1,065	4,467	20,944
1975	8,688	353	2,488	11,529
1976	4,769	1,176	2,488	8,676
1977	3,843	26	9,258	13,127
1978	8,421	229	3,663	12,313
1979	21,568	253	5,796	27,617
1980	17,084	2,203	3,175	22,462
1981	11,266	1,767	7,033	20,066
1982	10,502	3,550	10,796	24,848
1983	23,010	5,466	7,923	36,399
1984	22,103	6,104	5,188	33,395
1985	14,356	4,780	206	19,342
1986	131,949	6,513	5,369	143,831
1987	179,848	26,377	15,383	221,708

TABLE 4.3   Cables Restored by INTELSAT in 1987

Failed Cable	Terminal Countries	Restored	
		Circuits	Days
TAT-6	United States, France	3,368	22
TAT-7	United States, France	3,207	5
CANTAT-2	Canada, United Kingdom	1,570	11
SAT-1	Portugal, South Africa	330	20
Scotice	United Kingdom, Iceland	12	21
Canber	Canada, Bermuda	236	60
BER-1	United States, Bermuda	79	68
Atlantis	Brazil, Senegal	24	6
SEA-ME-WE	Singapore, Sri Lanka	678	13
Bermuda-Tortola	Bermuda, Tortola	63	32
Japan-China	Japan, China	72	89
United States-Jamaica	United State, Jamaica	142	6
Transpac-1	United States, Japan	130	11
TaiLu	Singapore, Taiwan	56	16
Hawaii-2	United States, Hawaii	139	19
East China Sea	Japan, China	60	26
Asean	Thailand, Singapore	252	10
ANZCAN	Australia, Canada	839	14

yet exceeded 3000 circuits. Restoring failed cables while they are repaired has, to date, been both practical and profitable for INTELSAT whose services are available for cables with capacities up to about 4000 circuits using transponders on a temporary lease basis.

In 1986, in order to accommodate an imminent failure of the TAT-6 cable, a Wideband Cable Restoration service was introduced by INTELSAT permitting cable owners to lease 36 MHz segments on hemispheric or zone-beam transponders on an annual basis and allowing the cable owners to use the capacity during the failure or planned maintenance of cables. The lease allows the capacity to be used for either 10 or 15 days each year and the rate for this 36 MHz circuit is US$12,000 to 13,000 per day. If the capacity is used for more than 30 days a year, the charge is reduced to about US$6000 per day.

The next generation of submarine cables, which start working in 1988–89 (TAT-8 and PTAT-1), and each fiber pair will have capacity of about 140 Mbps each way. Some planned cables are shown in Table 4.4. Capacities for the different cables are quoted differently by different owners and are confusing to interpret. However, they form a major step forward in submarine cable technology which returns the cable versus satellite status to that existing in the mid-1960s when TAT-3 and TAT-4 were laid each having capacity for 138 circuits compared with INTELSAT I and II satellites which were each able to handle about 240 circuits.

TABLE 4.4   Some Planned Fiber-Optic Cables

Year	Cable Name	Owners Between Countries	Fiber Pairs	Capacity Mbps
1989	TAT-8	AT&T and European PTTs U.S.A.-U.K.-France	3	3 × 280
1989	PTAT-1	Cable & Wireless and Tele-Optic (U.S.A.) U.S.A./Bermuda-U.K./Ireland	4	4 × 420
1990	TransPac1	AT&T and eight PTTs U.S.A.-Hawaii-Japan	3	3 × 280
1990	TAV-1	Submarine Light Cable U.S.A.-Europe	3	144 TV
1990	PACTEL1	Pacific Telecom Cable & Wireless U.S.A.-Japan	3	3 × 420
1990	H-J-K	Cable & Wireless, Japan KDD and Korea KTA Hong Kong-Japan-Korea	3	3 × 280
1991	PTAT-2	Cable & Wireless and Tele-Optic	4	4 × 420
1991	TAT-9	AT&T and European PTTs	3	3 × 565
1991	TASMAN2	OTCA and Australia New Zealand Post Office Australia-New Zealand		

The question of how INTELSAT satellites can be used to restore fiber optic cable systems has been under discussion at INTELAST for some years. In order to restore one of these new cables, such as PTAT-1, having a total capacity of around 1200 Mbps or over 750 T-1 circuits, in the event of their interruption necessitates readily available satellite capacity and a considerable amount of standby earth station capacity. A complete satellite may be needed as well as a standby earth station at each head of the cable. INTELSAT in its satellite services cannot be expected to make such capacity readily available unless the capacity is earning revenue in other ways or is paid for in its standby capacity. The cable owners require capacity quickly when it is needed, on a single satellite, and obviously at minimum cost. Late in 1987 this problem had still not been resolved.

The technical and economic factors have been made even more complex by the diverging interests of the various INTELSAT signatories which fall into roughly three categories:

Signatories with no cable interests
Signatories who also own cable shares
Signatories who compete with cable interests

Most countries are now beginning to have financial interests in cable systems as these are now expanding round the globe as they did about 100 years ago; the difference now is that they are mainly owned on a cooperative international basis rather than by private companies. European countries all use the major north Atlantic cables and thus cables and satellites are complementary to each other and not competitive. Many countries have no access to submarine cable systems and no interest beyond keeping down the cost of satellite use. The U.S. signatory stands out in a singular capacity as its primary business is to compete with cables because it is, as created by the Communications Satellite Act, only involved in satellites. COMSAT will, however, have a small investment in TAT-9.

With these very divergent views, cable restoration for fiber optic cables is not likely to be quickly or satisfactorily resolved.

4.5. EMERGENCIES AND DISASTERS

The characteristics of satellite communications make this medium ideal for providing emergency communications at times of national emergency or natural disasters. This was recognized by the World Administrative Radio Conferences in Geneva, 1979, which recommended that "administrations, individually or in collaboration, provide for the needs of possible relief operations in planning their space radiocommunication systems and identify for this purpose preferred radio-frequency channels and facilities which could quickly be made available for relief operations." INTELSAT, then and now, could be used to provide this service commercially on an occasional use basis using spare transponder capacity with small transportable earth stations. Other domestic and regional satellites can be used for their own region of coverage but these cannot offer worldwide coverage. INTELSAT did give consideration to providing the service and providing the means by making available some small earth stations which could be sent by air at short notice to anywhere in the world, but it would not have been commercial and users were not prepared to pay the cost. INTELSAT and INMARSAT, however, always have and probably always will provide any space segment capacity they can at any time and to any country in the event of major disasters or emergencies. INTELSAT now has two 1.8 m portable earth stations available for such emergencies.

A typical use of satellite communications at times of disasters was the Mexico City earthquake on 20 September 1985 when 10,000 people were feared dead and hundreds of buildings collapsed. The city was decimated and telecommunications paralysed by severed cables and damaged equipment which not only impeded rescue work but also broke off contact with the outside world, which was eager to assist. It is ironic that it is the outside world or news media that usually arrives with the emergency telecommunications equipment and Mexico was no exception.

Following almost total disruption of all radio, television, and telephone communications, it is usually amateur radio operators who raise the alarm that a disaster has occurred and provide basic telecommunications until some terrestrial means are restored, or until emergency satellite communications are made available through the nearest earth station—usually an earth station using INTELSAT or INMARSAT satellites—Mexico was no exception.

Mexico City was more fortunate than many similar disasters in that the Mexican domestic satellite, Morelos, had been launched successfully only a few months beforehand and which included totally unused Ku-band capacity. Although Mexico's main international gateway earth station at Tulancingo remained operational through the earthquake, almost all connections to it were severed and Mexico City's domestic earth station using Morelos was destroyed. It was the news media that provided the first major telecommunications facility.

The United States' NBC television network quickly contracted with COMSAT General to use its new satellite broadcast vehicle called Skybridge which was a mobile Ku-band earth station and transmitting facility. The Skybridge terminal was flown from Washington's Dulles Airport to Mexico City on the day after the earthquake aboard a Hercules C-130 cargo plane and then driven to the Juarez hospital which was the center of the rescue effort, and set up among the rubble of surrounding buildings.

This earth station provided NBC with a news scoop for television via Mexico's Morelos satellite to an earth station in Burbank, California, which then retransmitted it via the United States' SBS satellite nationwide, and via INTELSAT to the rest of the world. This earth station allowed other countries to see quickly the extent and magnitude of the disaster, the type of relief needed, and obtain worldwide assistance.

As well as the Skybridge terminal, COMSAT General also provided two suitcase earth stations (TCS-9000 type) used with INMARSAT satellites. These two terminals were installed at Mexico City's Channel 13 television studios and provided additional radio braodcast facilities for journalists.

The Associated Press (AP) similarly shipped earth station facilities and installed theirs a few days later at the Bank of America building. Other receive-only terminals were installed at AP subscribers' locations in and around the city for news and emergency information distribution.

Some political emergencies can be expected regularly and for this reason the United Nations, in 1982, sought assistance from INTELSAT towards the development of a United Nations lease for its peace-keeping operations, mainly in the Middle East. INTELSAT's Board approved such a lease in 1983 on the same basis as if it were a domestic lease, but with the proviso that it be used solely for peace-keeping and emergency relief activities. INTELSAT assisted the United Nations in developing a suitable network and planning its earth station in New York. The United Nations previously had earth stations in Jerusalem, Lebanon, and Switzerland that had been using the French Symphonie satellite until it reached the end of its life. These earth stations needed to be modified to work with INTELSAT satellites until an integrated new network could be installed, so INTELSAT provided free consulting services under its Assistance & Development Program to advise on modifications, specifications, and tender appraisals. The United Nations started its small 9 MHz lease in 1984 and now links the New York headquarters with emergency locations in Israel, Lebanon, and Switzerland.

In summary, many organizations, national and international, have suggested feasible satellite systems to help during major disasters and emergencies. When a disaster or emergency happens, usually all or any available resources are brought in as quickly as possible to assist, including communications satellites and earth stations. This means that every disaster creates a communications emergency and even after nearly a quarter of a century of communications satellites there is still not a routine procedure for handling it; every disaster is still handled on an ad hoc basis. The technology and equipment has been available for well over a decade, yet the world's largest satellite organization, INTELSAT, still has not been able to introduce a standard service that could be used by any country at any time. This situation is changing somewhat as INTELSAT now has its two 1.8 m portable earth stations that can be carried as hand baggage on commercial airlines to cover this type of emergency and on a strictly commercial basis.

4.6. HUMANITARIAN APPLICATIONS

INTELSAT has, since its inception, been involved with the telecommunications services necessary following emergencies and disasters whether they be mini-wars or earthquakes. INTELSAT has similarly been involved in a number of health and educational projects for many years, providing periods of free satellite use for such purposes as well as providing leased transponder services to the United Nations for its peace-keeping operations.

The first major humanitarian application was Project Hope in 1973 when a small terminal was placed on the hospital ship Hope anchored off Maceio in Brazil by COMSAT, the U.S. signatory, and showed how satellites could be used as a complete teaching and operating medical facility connected to a hospital on shore. Since then, INTELSAT has continued to provide free use of its satellites for test and demonstration purposes. Project Share reemphasized this between 1984 and 1987 during which it encouraged use of satellites to develop education and health care projects.

One of the last and most successful Share projects was a series of medical education broadcasts from the Miami Children's Hospital to 5000 doctors in Chile, Colombia, Peru, and other Latin American countries, which promoted and educated so many of the most needy child medical units on the continent.

A new program, Project Access, was started in December 1987 to succeed the successful Project Share, which will continue to provide free satellite use for educational, health, or other closely related social service.

5

BUSINESS APPLICATIONS

The use of digital computers followed by the deregulation of the tele-communications industry within the United States in the late 1970s and early 1980s generated an enormous surge in the requirements for private business networks which could most effectively be met only by satellite communications. This gave birth to a requirement now known as an integrated system digital network (ISDN) which would cover all business requirements such as telephone, data, facsimile, and video. Domestically, within the United States, the satellites belonging to RCA, Western Union, COMSAT, AT&T, and Amsat were quickly planned to satisfy these business needs, even though they were not known then as being similar to ISDN.

The question of introducing a range of new international digital services was first raised by INTELSAT's own staff in mid-1981. Following a study of possible requirements, satellite availability, and political aspects, the IN-TELSAT Business Service (IBS) was approved for use on a worldwide basis in December 1983, and the first IBS service started almost immediately between its first users—the Bank of Montreal's head office in Canada and its branch office in London, England. Since that time the growth of international, as well as national, satellite business networks has been impressive.

Satellite business networks were designed to permit businesses to bypass all the problems associated with long terrestrial networks. This bypass was seen also as a large cost saving by circumventing the large, and often monopoly, telecommunications carriers, both domestically and internationally. This bypass of terrestrial facilities is shown in Figure 5.1. Despite the enormous advantages to businesses for using such a network, especially where large companies were concerned, it was not surprising that there was reluctance by some INTELSAT signatories to encourage use of this new service internationally. This reluctance has now almost completely disappeared and satellite business networks, large and small, and here to stay and grow nationally, regionally, and internationally.

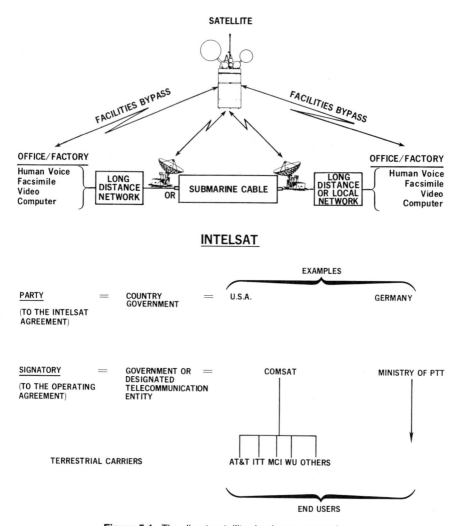

Figure 5.1. The direct satellite–business connection.

5.1. SATELLITE BUSINESS NETWORKS IN THE UNITED STATES

The first satellites with this purpose in view were launched by Satellite Business Systems (SBS) in 1980, and the American Satellite Corporation (Amsat) which eventually launched its first satellite in 1985 after using other operators' satellites for several years. SBS satellites were specifically designed for the new market, and the company was initially a joint venture between COMSAT General, International Business Machines (IBM) and the AETNA Insurance company. The second, Amsat, was initially an offshoot of Fairchild Industries and Continental Telephone.

The purpose of both companies was to provide a full range of services for large business users. The services envisaged were conventional voice and data communications as well as high speed data links for computer communications and also video services for video-conferences. The growth of these services was not as quick as had been forecast, and the first years for SBS were lean. However, from only three corporate networks using satellites for leased circuits in 1982, the total grew to 25 by 1985, according to a report issued by the NTIA. A survey by the magazine *Satellite Communications* in May 1986 listed 99 private satellite networks. The corporations using satellites for corporate networks included:

Hewlett-Packard	K-Mart
Chrysler	MicroAge Computers
Ford Motor Company	Walmart
J.C. Penney	Electronic Data Systems
Eastman Kodak	E.F. Hutton
Texas Instruments	Holiday Inn
Merrill Lynch	Safeway Stores
MONY Financial Services	Computerland
Digital Equipment Corp.	McDonalds
Sears Roebuck	General Motors

In addition to these there were a number of new companies that specialized in providing videoconference facilities on an occasional use basis, and which provided all the studio and setup arrangements. Other companies were formed that specialized in electronic mail, facsimile, and small computer links for the collection and distribution of data to or from multiple access points. Some of these companies were:

Videoconferences

Videostar Communications
Private Satellite Network (PSN)
Holiday Inn
World Port (Visnews and Western Union)

Electronic mail and facsimile

Federal Express

Computer networks

Equatorial Communications (now part of CONTEL)
Computer Science Corporation

Vitalink
Private Satellite Network (PSN)
Reuters
COMSAT
Telenet (US Sprint)

The most successful satellite networks that emerged were the simpler ones using proven equipment rather than those based on some of the advanced, but unproved, digital technologies which took several more years before their success was proved.

For small requirements, a business does not normally want to go to the additional expense of having its own earth station, and finds it more cost effective to use the terrestrial carrier tariffs either on an occasional use or a full-time leased basis. In the case of businesses with higher requirements, however, the economic and technical advantages of having its own earth station and bypassing both the cost and sometimes unreliability of the terrestrial network, the choice is quite different.

The U.S. customer was becoming accustomed to seeing small earth stations located on the premises of larger corporations, hotels, local television stations, and even in the private gardens of houses. Several manufacturers were successfully marketing small earth stations at quite low prices, and it was clear that the same sort of facility would eventually be required for international purposes as well.

Carrier bypass services such as IBS were seen by several large companies as a new and improved way of pursuing business internationally as well as within the United States. Some other countries saw it as a potentially dangerous service that would erode revenues. Later, some countries realized that the bypass was inevitable and began to see IBS as a new source of business, such as leasing facilities and maintenance contracts as well as freeing terrestrial lines for other purposes. Most signatories are now actively encouraging the IBS services on an international basis although in many countries there are still limitations on two-way satellite business services directly from customer premises.

Returning to the United States, many data network services have developed during the 1980s to accommodate the increasing need for companies to distribute and collect data from distant points. Such services can sometimes be obtained through the existing switched terrestrial network, but there are usually cost and technical benefits to be obtained from using satellites. A survey by Frost & Sullivan in June 1988 expects private sector satellite revenues to rise from US$567 million in 1987 to US$1.5 billion by 1992.

The earlier private satellite network suppliers such as SBS, COMSAT/AT&T/Comstar, General Telephone Electronics (GTE), and Western Union had already introduced full transponder services, but these were too large for many of the smaller needs, so new retailers of services emerged which would purchase or lease a transponder and sell smaller services to a larger number of customers.

Equatorial Communications (CONTEL/ASC). One of the first to develop this type of service, and also to pioneer the necessary earth station equipment was the Equatorial Communications Company of Mountain View in California in 1981, which is now a manufacturing subsidiary of the CONTEL Group. Equatorial networks provided end-to-end communications for smaller businesses, without any need for terrestrial circuits, by using its own microterminal earth stations with an antenna diameter of only 0.6 m. The satellite network operates at the same speed as for data communications but with improved performance characteristics as those typically carried over voice grade telephone lines. Economy of use is helped by offering the service with a network capacity in increments from 150 to 19,200 bps.

Equatorial's data networks were based on the use of spread spectrum technology which had been developed over 30 years of radar radio astronomy applications. Spread spectrum technology, which is patented in the United States and several other countries, minimizes interference from other satellite or terrestrial sources by using more frequency bandwidth than conventional signals transmitting at comparable data rates, but it permits much smaller receiving antennas.

Networks can be configured as a one-way distribution system, as a one-way collection network, or as an interactive network. Connection between user terminals is also possible through the host or master earth station. These configurations are shown in Figure 5.2 and a typical earth station block diagram is shown in Figure 5.3.

Equatorial purchased two satellite transponders on the WESTAR IV satellite and started to provide satellite packet switching networks within the United States. The equipment was later used on other satellite systems using C-band or Ku-band antennas; similar equipment is also used for a number of INTELSAT's INTELNET networks. Users of the Equatorial equipment or services include the Farmers Insurance Group with about 2500 sites, Reuters, Associated Press, Dow Jones, Merrill Lynch, the New York Stock Exchange and many others, both in the United States and abroad. The Farmers Group employs over 1400 people at its Los Angeles headquarters with some 14,000 agents nationwide. Their new network permits agents to communicate interactively with headquarters using IBM 5280 terminals. Other networks such as the news distribution services are mainly one-way services to the service subscribers.

K-Mart. K-Mart, which is the second largest retail store chain in the United States (Sears Roebuck is the largest), is constructing a US$40 to $50 million corporate satellite network to connect the chain's 2100 stores using US Sprint TELENET and Ku-band capacity on a GTE satellite. The K-Mart

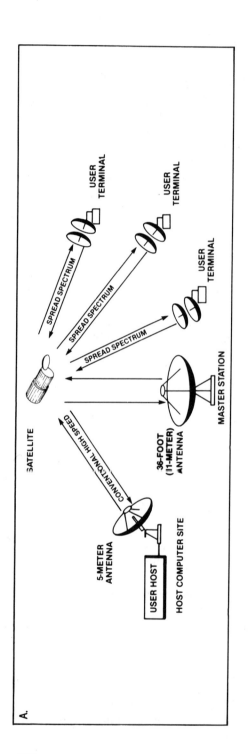

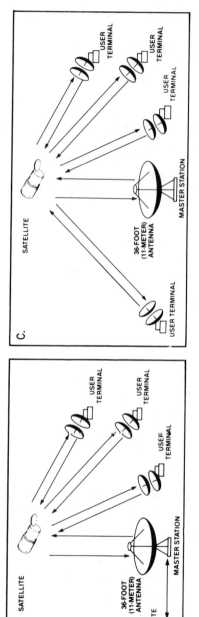

Figure 5.2. Spread spectrum techniques (courtesy: CONTEL/ASC. Protected by U.S. patent 4,455,651 and various foreign patents).

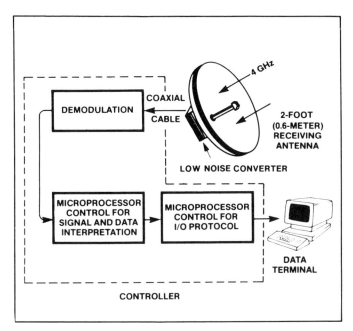

Figure 5.3. Spread spectrum earth station block diagram (courtesy: CONTEL/ASC. Protected by U.S. patent 4,455,651 and various foreign patents).

network is used for credit authorizations, inventory control, pricing, energy monitoring, and video demonstration applications. K-Mart preferred a fixed cost service that could be depreciated over the life of the system rather than a variable cost system using leased lines as well as saving about 60 percent compared with a leased terrestrial network. The K-Mart network will shortly link all its stores nationwide to its international headquarters in Troy, Michigan. The small antenna at each store, manufactured by NEC America, can usually be seen either on the roof or near the store's office facilities at the rear of the store. The network could extend overseas, if required, using the INTELSAT system with which GTE Spacenet is also connected.

Safeway Stores. Safeway Stores Inc. is a grocery and supermarket chain with nearly 2000 stores within the United States and more in the United Kingdom. Safeway started testing a satellite network for in-store advertising and broadcasting early in 1985, and its success prompted Safeway to extend the service later the same year. The programming of the service is contracted out to Instore Satellite Network which is a division of Broadcast International Inc.; they provide and maintain the earth station equipment at all the stores and also produce the programs; Safeway just has to dictate the type of programs required. The service provides data, audio, and video for introducing new products and also for management training. Broadcasting is done from studios in Salt Lake City and Fort Worth via a GTE Spacenet satellite.

Hewlett-Packard. Hewlett-Packard is another corporation that has combined video and data services into a private satellite network to improve its marketing, sales, and training activities. The Hewlett-Packard satellite network started in 1984; prior to this, lines were leased from AT&T's Digital Dataphone Service (DDS) each of which was capable of handling 9.6 kbps. The transmission rates were insufficient for transmitting large files of data which had to be shipped round the country by courier. A review of the company requirements showed that they needed a private network, free of price fluctuations, and able to handle full motion videoconferencing (T-1, 1544 Mbps) as well as 56 kbps data channels.

It became clear that no terrestrial system at that time could meet these requirements. It was also found that while a satellite system would be US$868,000 more expensive to install, it would cost US$500,000 less each year to operate and that a break-even point would occur in about two years. Vitalink earth station equipment was selected to link the three main offices at Fort Collins (Texas), Boise (Idaho), and Cupertino (Colorado). The earth stations have 9.2 m diameter antennas with full backup equipment against failure and to permit maintenance. The network uses two C-band transponders on the AT&T Telestar satellite but operational, technical, and management functions are performed by Vitalink through its operations center at Mountain View in California.

Public Switched 56 kbps Services. AT&T almost introduced a switched 56 kbps service known as called switch digital capability (CSDC) prior to divestiture in 1983, but this had to be postponed until early 1985 due to the change of ownership of the local phone companies. A new service called Accunet was introduced in 34 major cities which has grown steadily and permits a wide variety of new customer applications on a dial-up basis. Since divestiture, a variety of competitive public switched data networks have emerged in the United States.

Northern Telecommunications formed a similar new Datapath service using its own equipment but it is incompatible with the AT&T equipment. GTE Spacenet introduced a new service called Call Express which joined up its very small earth station network into its public switched system at its McLean (Virginia) headquarters control, using Skyswitch small earth station equipment. MCI introduced its Mail service for similar digital switched services.

Comsat Technology Products Inc. (which was then a COMSAT subsidiary, but was later sold to CONTEL) introduced its Starcom public switched data network in September 1986 for 56 kbps shared networks, and the number of similar services continues to increase, promoting many competitive business services of this nature. Almost all of these new services are using very small earth stations (commonly called VSATs) and just like the number of services, the choice of earth station equipment is proliferating quickly.

5.2. BUSINESS NETWORKS IN EUROPE

A similar wide variety of satellite business service has also emerged in Europe using EUTELSAT's Satellite Multi Service (SMS) which is equivalent to, and compatible with, INTELSAT's IBS. Both of these services are verging towards the future ISDN network. SMS is available on transponders 11 and 12 of the EUTELSAT satellite, or on France's TELECOM I satellite under an agreement between EUTELSAT and the French PTT.

Business networks in Europe are almost universally provided through each country's telecommunications administration and lack the competitive influence that has been introduced in the United States. The United Kingdom introduced some degree of competition in its telecommunications when a private company, Mercury Communications, was granted authority to provide telecommunications services in competition with the then Government-owned British Telecommunications. Even this limited competition is still the exception rather than the rule in Europe.

European industry is now manufacturing its own small earth station equipment and this is helping the further development of small business networks, both within Europe and between Europe and other continents.

5.3. INTERNATIONAL BUSINESS NETWORK SERVICES

Up to the present time all international business network services have been provided by submarine cables or the INTELSAT satellites. Many leased international business networks have used, and continue to use, terrestrial cable systems but predominantly they use satellites of the INTELSAT system where intercontinental or global networks are involved with multipoint connections.

5.3.1. The INTELSAT Business Service (IBS)

This service was first approved in December 1983 as a totally digital integrated service forerunner to ISDN and was designed to provide a full range of business services directly, or indirectly, to business premises. It is not permitted, however, to be used for switched public services but the types of application include:

Video and audio teleconferencing
Facsimile

Data transfer

Computer-aided design

Digital voice and data

Remote printing of newspapers

Electronic mail

Data collection

Data distribution

Electronic funds transfer

Audio program distribution

These applications are far from comprehensive, as any digital application can be serviced through customer premise earth stations, or teleports, as well as through the large national gateways and the principle telecommunications carriers.

The services are offered by INTELSAT to its signatories with a wide variety of terms and conditions which are available on a nonpreemptible basis as a full-time lease with a minimum of three months, or as a part-time lease with a minimum of one hour with increments of 15 minutes. The minimum lease period of three months carries the option to renew, which encourages service trials. Occasional use service is available on an as-needed basis with reservations for periods of 15 minutes in incremental units. There are cheap rates for part-time and occasional use services during off-peak such as between 20.00 and 24.00 GMT in the Atlantic satellites when normal traffic is at its lowest.

Various businesses have widely differing requirements with regard to the data rate needed, and INTELSAT's service takes this into account covering bit-streams from 64 kbps to 8.4 Mbps. Low-medium capacity bit-stream rates of 64, 128, 256, 384, 512, and 768 kbps are offered which are suitable for such applications as low-medium speed data transfer, facsimile, and digital voice. High capacity bit-stream rates of 1.544 to 8.448 Mbps are available for such applications as full motion, full colour video services, high speed data transfer, or multiplexed services.

Full or fractional transponder leases are also available in 9 MHz increments up to 72 MHz for a full transponder, and it is now possible to obtain a bit stream of 140 Mbps using a 72 MHz transponder.

Late in 1987, 17 countries were using INTELSAT's IBS service at bit rates mainly up to the T-1, 1.544 Mbps rate. This service, although slow to start, is now growing exremely fast as shown in Table 5.1.

The quality of the service is high. Basic IBS provides high reliability and channel availability. Under clear sky conditions and nominal bit-error-rate (BER) of 1 in 10 to −8 or better is achieved through use of 1/2 forward error connection (FEC). Under poor sky conditions, such as rain, a BER of 1 in 10 to −6 for 99 percent of the time is provided assuming the same FEC code rate. Individual networks can be equipped with different FEC systems

TABLE 5.1   Growth of INTELSAT's Business Service (IBS)

Year	Region			Total
	Atlantic	Indian	Pacific	
Countries Using the Service				
1984	2	–	–	2
1985	4	–	–	4
1986	11	–	–	11
1987	14	2	6	22
1988	19	6	6	31
Growth in 64 kbps Units				
1984	16	–	–	16
1985	296	–	–	296
1986	1,556	–	–	1,556
1987	3,146	2	498	3,646
1988	5,370	610	1,300	7,280

to obtain an even better link performance to suit special user needs or overcome local rain conditions. A super IBS service was introduced in 1986 for the few businesses that require an even higher performance over a longer period of time.

QPSK/FDMA using either SCPC or multichannel per carrier techniques are used but INTELSAT will consider TDMA requirements on a case-by-case basis.

Earth station standards for INTELSAT's IBS are shown in Table 5.2 and permit earth station antennas from about 3 m upwards.

The IBS can be used for what are often termed open or closed networks. Open networks allow the users to interconnect their own equipment accord-

TABLE 5.2   INTELSAT IBS Earth Station Standards

Standard	Antenna Size (m)	Frequency Up/Down (GHz)	G/T		Rate Adjustment Factor	
			Band	dB/K	Basic	Super
A	23–32	6/4	C	35	1.0	N/A
B	10–13	6/4	C	31.7	1.12	N/A
C	17–32	14/11	Ku	39	1.0	1.50
E-1	3–5	14/11	Ku	25	1.87	4.15
E-2	5–6	14/11	Ku	29	1.37	2.47
E-3	8–10	14/11	Ku	34	1.12	1.87
F-1	4–5	6/4	C	22.7	2.24	N/A
F-2	7–8	6/4	C	27	1.49	N/A
F-3	9–10	6/4	C	29	1.25	N/A

ing to agreed parameters and performance standards. Closed networks are not so closely defined and allow individual users to choose a network suitable to their specific needs. For closed networks there are fewer constraints and performance requirements which make them easier to assemble; almost all networks thus far are closed networks for this reason.

Three network concepts allow each user to choose the most suitable network for its own needs as shown in Figure 5.4. A User Gateway uses small earth stations such as Standard E1, E2, F1, or F2 which can be used on the user's premises with very short interfaces to the user's own computer or other facilities. One such earth station can serve a single user or be shared by a number of small users.

The Teleport Gateway goes one stage further where a group of earth stations are located near major cities. Teleports usually have earth stations accessing a number of different satellites to provide the city's business community with a wide variety of telecommunications services, both domestic and international. The number of teleports is growing rapidly, especially in the United States, and they are linked to their respective customers by fiber optic cables or direct microwave links to minimize the length and reliability problems associated with terrestrial links.

The Country Gateway concept uses a large, usually existing, earth station and the country's terrestrial network to connect the digital international service to the customer.

In order to accommodate the very varied needs for such a service, various satellite beam connections are possible with three fundamental types of connectivity which are shown in Figure 5.5. The Basic Connectivity Service

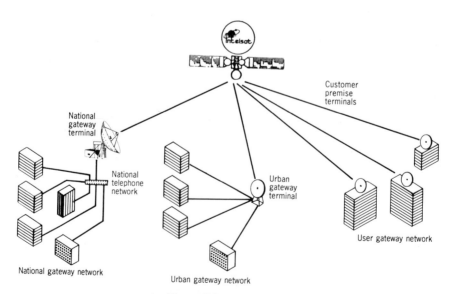

Figure 5.4. IBS gateways.

is a one-way communications path with a single down-link beam which can be west-to-east, east-to-west, east-to-west, east-to-east, or west-to-west from the transmitting station to the receiving station. The path may use either two C-band beams or two Ku-band beams provided that:

All receiving points are in the same down-link beam

All sending points are in the same up-link beam

All sending points time-share the same channel

Enhanced Connectivity Service, which costs 1.5 times the basic connectivity charge, is a one-way communications path with a single down-link beam which again may be west-to-east, east-to-west, east-to-east, or west-to-west, but provides a connection (commonly called cross-strap) in the satellite, between the Ku-band and C-band beams provided that:

All receiving points are in the same down-link beam

All sending points are in the same up-link beam

All sending stations time-share the same channel

Full Connectivity Service became available with the INTELSAT V-A satellites and provides multiple beam interconnection in the satellite so that a transmission on one hemisperic or spot beam (east or west) can be received simultaneously in east and west hemispheric and spot beams at C-band and at Ku-band. All sending points time-share the same channel, and rate adjustment factors are applied to the number of down-link beams used (1.0 for one beam; 1.8 for two beams; 2.7 for 3 beams; and 3.4 for four beams).

All these services allow use of earth stations on user premises which, depending on each country's regulations, may be user-owned and user-operated with the user having direct access to INTELSAT satellites. The INTELSAT Agreements are similar to an international treaty, and it is each country's signatory that is responsible to INTELSAT for the operations of earth stations under its control. It is for this reason that the signatory to INTELSAT must:

Organize the operational guidelines for the earth stations within the scope of all technical and operational confines layed down by INTELSAT,

Distribute and ensure adherence to all INTELSAT characteristics and guidelines such as mandatory technical characteristics and guidelines such as those recommended in INTELSAT's Satellite System Operational Guide (SSOG),

Be responsibe for all contacts with INTELSAT, payments to INTELSAT for the services used, and

Create its own institutional arrangements covering IBS tariffs to conform with the country's particular requirements.

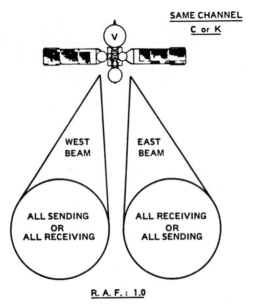

SAME CHANNEL
C or K

WEST BEAM

EAST BEAM

ALL SENDING
OR
ALL RECEIVING

ALL RECEIVING
OR
ALL SENDING

R. A. F. : 1.0

I. B. S. BASIC CONNECTIVITY
East to East, West to West
East to West or West to East

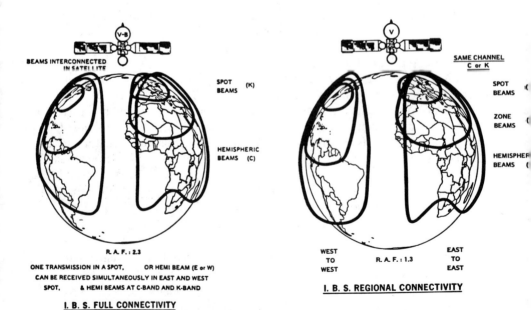

BEAMS INTERCONNECTED
IN SATELLITE

SPOT BEAMS (K)

HEMISPHERIC BEAMS (C)

R. A. F. : 2.3

ONE TRANSMISSION IN A SPOT,   OR HEMI BEAM (E or W)
CAN BE RECEIVED SIMULTANEOUSLY IN EAST AND WEST
SPOT,   & HEMI BEAMS AT C-BAND AND K-BAND

I. B. S. FULL CONNECTIVITY

SAME CHANNEL
C or K

SPOT BEAMS

ZONE BEAMS

HEMISPHER BEAMS

WEST
TO
WEST   R. A. F. : 1.3   EAST
TO
EAST

I. B. S. REGIONAL CONNECTIVITY

Figure 5.5. Types of IBS connectivity (courtesy: INTELSAT).

The space segment charges are levied by INTELSAT to its signatories for the use of the satellite; they are NOT the charges made by the signatory to the end-user, as they do not include the cost for the earth stations or any terrestrial equipment, nor do they include any other signatory services that may be provided. It is common, however, for end-users to reduce costs by 50 percent by using IBS for large business services rather than long terrestrial links.

IBS is available on a number of INTELSAT satellites covering all three ocean regions and INTELSAT has committed itself to making more capacity available if and when the need arises. There was some delay in introducing full connectivity services due to the launch failure of the first modified V-A satellite F-14 in May 1986, but now that the INTELSAT V-A, F-13, which was launched successfully in May 1988, is satisfactorily operating at 307°E, full services are possible. The modified V-As have enlarged Ku-band spot beams which provide considerably more land coverage of the United States, and the satellite located at 307°E can provide IBS between Europe and the whole of the contiguous United States. Services are also obtainable on the satellite located over the Atlantic ocean at 342.5°E and over the Indian and Pacific ocean satellites.

Since the beginning in 1984, IBS has been provided (or planned) for U.S businesses by a multitude of carriers including:

AT&T
ALLNET Communications
COMSAT International Communications
CONTEL/AMSAT
Cresscomm Transmission Services
FSM Telecommunications Corp.
Fairchild Industries
Fedex International Transmission Corp.
FTC Communications
GET Spacenet
Hawaiian Telephone Company
Hughes Networks
MCI International
IDB Communications
ITT World Communications
International Relay Inc. (IRI)
NET Express
Puerto Rican Telephone Company
RCA Global Communications
TRT Communications (now Pacific Telecom)

OrionNet
Overseas Telecommunications Inc. (OTI)
National Gateway Inc.
U.S. Sprint
Vitalink Communications
FTC Communications
Western Union Worldcom
World Communications

Teleports

Houston International Teleport
South Star Communications
Boston Teleport
Turner Teleport International
Washington International Teleport
Manhattan Teleport

and the list continues to grow.

IBS Network Applications

Bank of Montreal. The Bank of Montreal was the first multinational corporation to use the new INTELSAT Business Service using a small 4.5 m antenna mounted on top of the 72 story First Canadian Place Building in Toronto. INTELSAT's Canadian Signatory, Teleglobe Canada, provided the earth station facilities in Canada and the United Kingdom signatory, British Telecommunications International, provided the earth station and terrestrial facilities in the United Kingdom. The private services in this network include voice, facsimile, electronic mail, audio plus graphic teleconferencing, and funds transfer in the initial stage, but which will be extended to full videoconferencing.

Massey Ferguson. Massey Ferguson is another Canadian business which has facilities throughout the world and is multinational in structure. It has manufacturing plants in Canada, United Kingdom, France, and Italy with its world headquarters in Toronto, Canada. These units need more and more computer communications between them as well as electronic mail and other office services. Data centers existed in both Canada and the United Kingdom but it was found that a much greater transatlantic capacity was needed, and that satellite transmission would prove to be a much more cost-effective means of communication as the needs increased. Massey Ferguson decided to plan along the satellite line starting in 1983, and followed the lines of the Bank of Montreal using some common facilities that are provided by Teleglobe Canada. Both these Canadian applications are shown diagramatically in Figure 5.6.

Morgan Guaranty Bank (MGB). The Morgan Guaranty's main office is in New York City, and the bank required improved communications with its office in London, England. MGB selected FTCC McDonnell Douglas International Communications and British Telecommunications International to cater for their IBS needs on the respective sides of the ocean. The

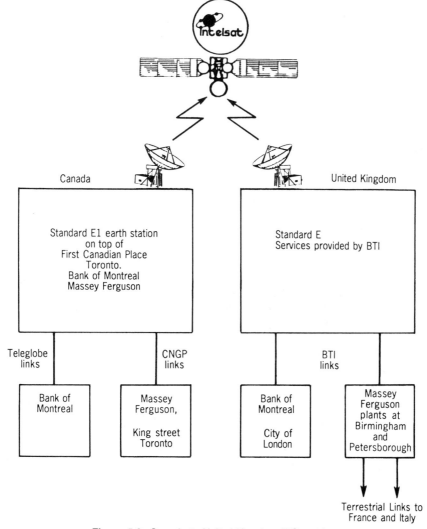

Figure 5.6. Canada to United Kingdom IBS services.

services first covered voice and data, but expanded to include facsimile a year later.

FTCC's earth station facilities are located at the New York Teleport on Staten Island together with the associated fiber optic cable system for the service to the main office on Manhatten Island. In London, the service passes through BTI's earth station and the Satstream digital services network to MGB's London office.

XEROX Corporation. Xerox required a 768 kbps data stream intially but with the later capabilty of expansion to a full T-1, 1544 Mbps capacity for its digital services. Xerox chose the American Satellite Company (now CONTEL) which provides capacity for their domestic network within the United States, to provide a Standard E2 earth station at their headquarters in Rochester, New York. In London, Xerox chose BTI to provide the terrestrial tail using BTI's Ealing earth station and the Satstream digital network to their London office. Xerox believed that the IBS would provide the most flexible and cost-effective solution for the company's network requirements, which will eventually include Japan and Brazil. Xerox's domestic and international networks are shown in Figure 5.7.

Printing European Newspapers in the United States. The international edition of London's Financial Times has been available in the United States since 1979 but distribution could only be handled by airfreight on a daily basis from the papers printing plant in Frankfurt, West Germany. On 12 June 1985 the Financial Times became one of the early users of the IBS in the United States using a network designed jointly by the American Satellite Company and British Telecommunications International to link the newspaper office in London directly with the printing facilities in Bellmawr, New Jersey.

During the first year of service the U.S. circulation jumped from 6000 to 10,000 copies each day. The network now uses 256 kbps to transmit the typeset version of the paper from the London headquarters to Frankfurt and New Jersey. It was decided not to have a rooftop antenna at the London office as the building, Bracken House, is not very high and constant construction could create visibility problems at any time, especially if it later become desirable to transmit via other satellites as well.

A very similar service is now also being provided between Paris and Miami for printing the International Herald Tribune in the United States. The International Herald Tribune uses the French Administration's digital terrestrial services in France, but at Miami they chose the Overseas Telecommunications Inc. to construct and operate earth station facilities at the printing plant.

IBS Network Costs. As mentioned earlier, there are now many companies retailing INTELSAT's Business Services within the United States. Some of

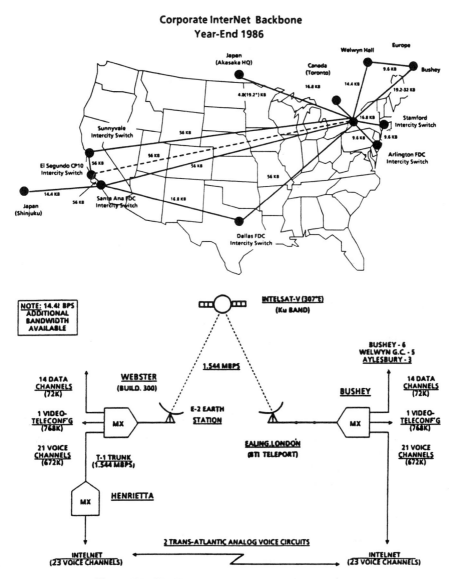

Figure 5.7. The Xerox business network (courtesy: Xerox).

these, notably AT&T, GTE, MCI, COMSAT, and CONTEL link INTEL-
SAT to their own satellite or fiber optic cable systems within the United
States. To reduce the terrestrial links to a minimum, these companies are
now building earth stations near major business centers round the country to
obtain direct access to INTELSAT. Many other companies provide the
earth station facilities to customer premises and lease the INTELSAT
service through COMSAT, the U.S. signatory to INTELSAT.

From the user's viewpoint, service offerings are becoming increasingly competitive and service costs have dropped rapidly. It is also useful for any potential user to analyze these costs to determine which are likely to remain constant and which are likely to fluctuate.

Considering a single T-1 (1.544 Mbps) data channel between the United States and the United Kingdom (or any two countries), there are three basic elements:

1. INTELSAT's space segment charge, which at the time of writing would be about US$90,000 per channel per year for each end of the circuit. This would be for the basic service between two Standard E2, 8-m earth stations, but if a smaller customer premise earth station were to be used, this charge would be multiplied by a factor of 1.33.

2. The signatory charge to the terrestrial carrier which would include the INTELSAT charge to the signatory, and the signatory's costs for managing its country's obligations to INTELSAT. At the time of writing in the United States, COMSAT's charge to the terrestrial carriers is about US$146,640 per year. If COMSAT also provides the earth station facilities the cost becomes about US$300,000. In the United Kingdom the same types of services are available through either the United Kingdom signatory, British Telecommunications International, or the one other company permitted to operate such services, Mercury Communications. Costs in the United Kingdom have also become more competitive since the deregulation in that country.

3. The carrier charge to the end-user, which is the real variant, and differs widely and includes different terms and conditions. They all include the INTELSAT charge and the signatory charges. At the time of writing, some typical service costs for a 1.544 Mbps T-1 channel, as widely reported in the trade press, are:

     AT&T's Skynet Service US$360,000 per year

     Overseas Telecommunications Inc. $258,000 per year

     BTI's Satstream Service $225,000 per year

     Mercury Communications $210,000 per year

Since deregulation, the the introduction of competition between the national carriers, there has been a marked tendency for tariffs to drop but it can be seen that the satellite cost represents only about 11 percent of the total cost to the service user and that any downward cost of the satellite unit is unlikely to have much affct of the user cost. The breakdown of service costs between the United States and the United Kingdom using costs quoted in the trade press at various times is shown in Figure 5.8 for the same 1.5444 Mbps T-1 service. Several companies offer variations of this service and costs are becoming increasingly complex and competitive.

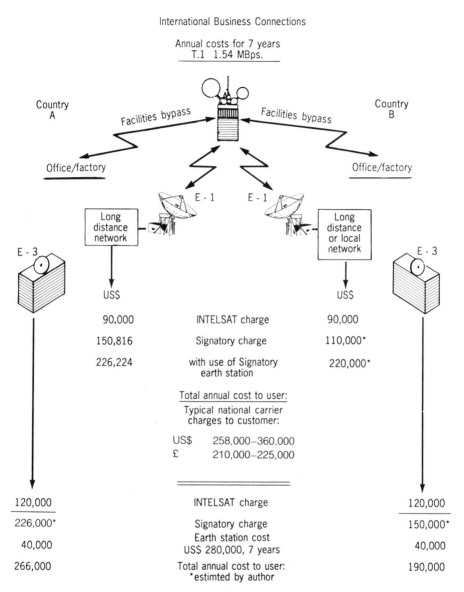

International Business Connections

Annual costs for 7 years
T.1 1.54 MBps.

	US$		US$
	90,000	INTELSAT charge	90,000
	150,816	Signatory charge	110,000*
	226,224	with use of Signatory earth station	220,000*

Total annual cost to user:
Typical national carrier
charges to customer:

US$ 258,000–360,000
£   210,000–225,000

120,000		INTELSAT charge	120,000
226,000*		Signatory charge	150,000*
40,000		Earth station cost US$ 280,000, 7 years	40,000
266,000		Total annual cost to user: *estimted by author	190,000

Figure 5.8. Typical 1.5 Mbps network costs using IBS.

User Earth Station Equipment Costs. If the user opts to install its own earth station, it will involve a high capital expenditure for the earth station as well as the signatory's annual charges for the satellite capacity. The present cost for a very adequately equipped earth station installed as a turnkey project ranges between about US$180,000 and US$250,000 depending on the equipment options. There will, of course, be annual operations and maintenance costs which are not included. Transatlantic IBS connection costs between the United States and the United Kingdom, which are probably somewhat lower in these two countries than many others, are shown in Figure 5.8 which indicate the clear cost savings to be derived from bypassing the expensive terrestrial facilities and having the earth station on the premises for networks faster than 64 kbps. Contractual costs for operating and maintaining the earth station are not included, but they can be considered fairly minimal compared to the cost of leasing the longer terrestrial network, especially where a long terrestrial tail is necessary.

5.3.2. INTELNET Data Services

Another wholly digital INTELSAT service is INTELNET which has the additional advantage that even smaller antennas can be used at the earth stations. INTELNET is designed to transmit and receive all types of data to and from very small earth stations and it is best used for collection and distribution networks such as oil exploration, financial networks, remote printing, and similar applications. One-way networks are termed INTEL-NET I, whereas interactive, two-way, networks are termed INTELNET II; these are particularly useful for banking, airline or hotel reservation, and similar networks. Two typical networks are shown in Figure 5.9 for INTEL-NET networks in the Far East provided by the Australian Overseas Telecommunications Authority and Cable and Wireless (Hong Kong) Ltd.

The INTELNET service is offered in both Ku- and C-bands, on a fractional or full transponder basis; the smallest lease is 1 MHz bandwidth spread spectrum transmission techniques and also Binary Phase Shift Keying (BPSK), with one-half FEC coding, but other means are possible and permitted, subject to case-by-case approval. The service is available on a range of INTELSAT satellites and on a preemptible or nonpreemptible basis, whereas IBS is only available on a nonpreemptible basis. Transmission plans required technical approval by INTELSAT to ensure that all appropriate technical criteria are complied with. Occasional use service is also available, but only on a preemptible basis. The service costs vary according to the length of lease, the capacity leased, and the type of satellite beam used. Like other INTELSAT services, charges can usually be obtained from national carriers or by reading the trade press.

First introduced by ITNELSAT in 1984, and based in many ways upon similar networks already working in the United States, INTELNET is now emerging as a useful service with worldwide applications. Among the early service applications are Australia's SATNET system which is both domestic and international and links a number of remote islands and business interests.

SATNET offers two-way digital communications to 1.8-m earth stations operating through the Melbourne central earth station. In the United Kingdom, Mercury Communications introduced the service for its IBM and

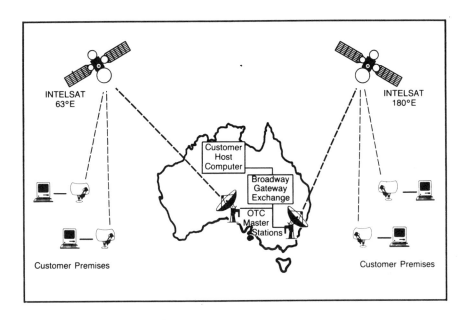

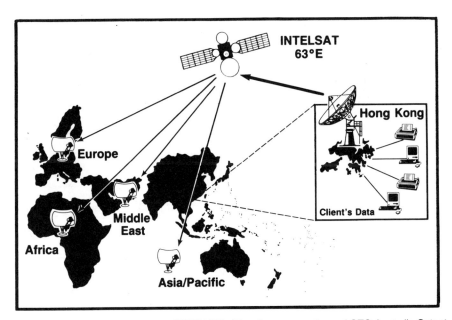

Figure 5.9. Typical data networks INTELNET. (Top diagram courtesy of OTC Australia Satnet Service; bottom diagram courtesy of Cable & Wireless (HK) Ltd.).

Electronics Data Service customers linking their British and overseas points. The Cable and Wireless Group has also introduced INTELNET as a one-way service in the Far East based upon their Hong Kong offices. This network uses a code division multiple access (CDMA) transmission system.

5.3.3. INTELSAT's VISTA Service

Although the VISTA service was created specifically for small rural communities in underdeveloped countries, the first application was very much business-oriented. The AMOCO Production Company set up an oil exploration headquarters at Madagascar in 1983 investing some US$80 million in the project. The project operated four concessions in western Madagascar

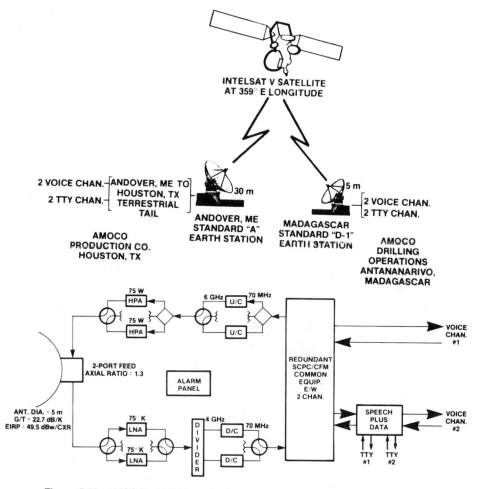

Figure 5.10. AMOCO's VISTA application for oil exploration (courtesy: AMOCO).

covering 10 million acres for which reliable two-way communications were essential between the field and AMOCO headquarters in Houston, Texas. Such communications did not exist and when AMOCO first sought assistance from INTELSAT in 1980, no suitable services existed although they were at a planning stage. It was May 1985 before it was possible for AMOCO to start service using the first of INTELSAT's new services, VISTA, which had been introduced in 1983.

AMOCO contracted with the GTE International Systems Corporation to provide an INTELSAT Standard D-1 earth station in Madagascar which could connect directly with the United States via an INTELSAT satellite located at 359°E and a COMSAT earth station in Andover, Maine. Terrestrial connections linked the Maine earth station with the company's headquarters in Texas. The network is shown in Figure 5.10. The station was used for 3 years before being moved elsewhere and used for providing IBS.

Although the AMOCO business application was the first practical use of VISTA, other examples of the service being used to link remote islands and small communities are emerging rapidly. Australia started the service to remote islands in 1985, and several countries are likely to introduce rural networks of this nature during the remaining years of the 1980s.

5.4. TELEPORTS

Teleports, or urban gateways, are groups of earth stations acting as a port for business interests in the same way as seaports or airports. The idea is that a teleport can provide the telecommunications needs of a specific business community without the problems and expenses of long terrestrial links. A teleport provides a common facility to provide earth stations to all the satellites that the business community needs, and can combine international, domestic, data, video, and all the satellite services needed by that community.

The deregulatory trend in the United States has encouraged a considerable number of new telecommunications companies to be formed to build such teleports, together with what have become known as local area networks (LANs). Not all of these provide international services but the number is growing rapidly both in the United States and other countries. One of the first, and probably one which will become the largest, is the Staten Island Teleport for New York City, and which is expandable to 350 acres. It will link all the earth station facilities of about 17 satellite antennas to the business of New York City by an underground fiber optic cable system offering voice, data, facsimile, video, and teleconferencing within the United States and to points overseas. The New York Teleport was a joint venture of the Port Authority of New York and New Jersey, and Teleport Communications Inc. which is a partnership of Western Union and Merrill Lynch Telecommunications Inc. It is already a large telecommunications complex and involves a very large investment for what is hoped will become a very profitable business. It is not surprising that other teleports have rapidly emerged in the same area, and even before it started working, an earth station belonging to International Relay Inc. (IRI) was providing IBS to some downtown New York City businesses. The New York Teleport obviates some of the terrestrial problems and costs that existed, but there remains the cost of the fiber optic cable system that is being laid between the teleport and the businesses. The New York Teleport was closely followed by other teleports offering international facilities in many other North American cities including:

California	The Bay Area Teleport
	Los Angeles Teleport
Florida	Miami Teleport
Georgia	Turner Teleport, Atlanta
Louisiana	New Orleans Teleport
Massachusetts	Boston Teleport
Texas	Houston International Teleport
Virginia	Washington International Teleport

In other countries, some of the new teleports existing or planned include:

Canada	Montreal Teleport
France	Paris, Rambouillet
Jamaica	Jamaica Digiport, Montego Bay
Mexico	Chihuahua
Netherlands	Amsterdam Teleport
United Kingdom	BTI's London Teleport
	Mercury Thameside Teleport
	Isle of Man Teleport

6

BROADCASTING APPLICATIONS

Television, first introduced in the United Kingdom by the British Broadcasting Corporation (BBC) in 1936, celebrated its 50th birthday in 1986 and "live by satellite" has become commonplace since transatlantic television broadcasting was inaugurated over INTELSAT I by the U.S. President Lyndon Johnson on 28 June 1965. Apart from entertainment, it has enormous uses in education which are slowly being implemented in the most needful parts of the world. It has largely replaced radio broadcasting as the world's information media.

In the case of satellite broadcasting, the differences between national and international becomes legally confused and technically almost impossible to control—just like Europe during World War II, without preventing the BBC from broadcasting news, people could only be prevented from listening by force. Today's TV is very similar and the first example of television distribution by satellite in the United States came the realization that it could also be received and redistributed in many surrounding countries. This allowed the unlimited reception of television programs. This situation has now appeared in Europe and elsewhere. As radio and television broadcasting use "wireless" or radio waves for distribution, there is no limit on any person receiving these broadcasts if they are receivable. The transmitter can apply coding techniques to prevent them being received by nonpaying recipients, but any earthbound satellite signal can be received, free of charge, within technical and legal limits.

Apart from the few existing regional satellite systems and the national satellite systems where programs are received by neighboring countries, INTELSAT was, until recently, the only real means for international television distribution. For many years INTELSAT has also been used by many countries for domestic television distribution. Combined, INTELSAT is the largest distributor of international television, but this could change when competing satellite systems emerge or when the U.S.S.R. introduces its TOR satellites providing higher power than INTELSAT, and which will provide worldwide coverage.

As we have already seen, the occasional use services have become so popular that the larger broadcasters have started to take up full-time television channels on an international basis. Television distribution by satellite has now become so widespread that during 1986, television programs were being transmitted through over 300 transponders using nearly 50 satellites (Appendix D). There are an increasing number of periodicals in North America and Europe that can provide up-to-date information on how these can be received, what frequencies are used, and what standards are employed.

The next step in television broadcasting will be when satellites can broadcast directly into individual homes which have very small receiving earth stations. This type of satellite has become known as the direct broadcast satellite (DBS). The DBS capability needs maximum power from the satellite to be transmitted for reception by the smallest possible earth station antenna. Such satellites have been talked about for many years and are now just beginning to provide services directly to homes in a few countries.

There were technical difficulties in developing DBS direct distribution as receive antennas must be very small for mounting on or inside, houses and the satellite power must be very high to achieve this. The power and earth station size problems have been overcome in recent years, but even so, the home earth station cost is still too high for many potential users, but what probably remains the largest problem, is for the transmitting investors to be able to ensure that all recipients pay for its reception; without this revenue the DBS investor is not recovering his very high investment in the satellite and programming costs.

The early market forecasts for DBS proved to be far too optimistic both in the domestic and international areas. They did not give full consideration to the political and social problems, and they overestimated the recipient's ability, or desire, to pay for the services. In the United States many potential DBS companies have not proceeded as planned. The European and Japanese DBS satellites, with their government-supported development and slow momentum, are the forerunners of DBS and will test the world market for value and usefulness. Japan, after a disappointing failure of its first DBS in orbit, launched its second DBS (BS2B) in 1987 and services are now said to be expanding quickly. However, there were no cable or satellite distribution systems in Japan prior to DBS.

6.1. INTERNATIONAL RADIO BROADCASTING

In the past, international radio broadcasting has meant shortwave radio transmission such as those beamed around the world by the Voice of America (VOA), the British Broadcasting Corporation (BBC), and Radio Moscow, to name just three. These programs are transmitted by powerful shortwave radio transmitters from the originating country and received by individuals around the world on any radio receiver equipped for receiving those frequencies. In many cases, these signals are received by foreign broadcasters and retransmitted locally. The latter course means that the listener gets a much superior quality reception than listening into the shortwave signal directly. It is surprising how many countires still transmit programs on the shortwave, which is an indication as to the demand for receiving news and programs from other countries.

Satellite communications services have long been used as the the intermediary between the original broadcasts and the local rebroadcaster. There

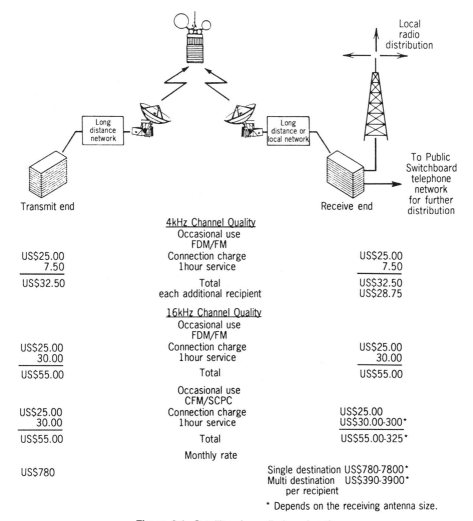

Transmit end		Receive end	
	4kHz Channel Quality Occasional use FDM/FM		
US$25.00 7.50	Connection charge 1 hour service	US$25.00 7.50	
US$32.50	Total each additional recipient	US$32.50 US$28.75	
	16kHz Channel Quality Occasional use FDM/FM		
US$25.00 30.00	Connection charge 1 hour service	US$25.00 30.00	
US$55.00	Total	US$55.00	
	Occasional use CFM/SCPC		
US$25.00 30.00	Connection charge 1 hour service	US$25.00 US$30.00-300*	
US$55.00	Total	US$55.00-325*	
	Monthly rate		
US$780		Single destination US$780-7800* Multi destination US$390-3900* per recipient	

* Depends on the receiving antenna size.

Figure 6.1. Satellites for radio broadcasting.

are two ways for providing the link by satellite; either using each country's telephone network and the satellite link to reach the rebroadcast station or to install small earth stations at each rebroadcast station as shown in Figure 6.1. Further distribution is sometimes obtained using public switched telephone networks. This method is used by France's Worldwide Sound Program Network through TELECOM-1 satellite and public telephone networks in Europe, Guadaloupe, Martinique, French Guiana, and Reunion. Satellites do not yet provide direct radio broadcasts to individual portable radio sets, and indeed such broadcasting is fraught with political as well as financial problems, and at the moment there are no satellites designed to have this capability. The first satellite that is likely to come close to this capability will be the Soviet Union's TOR satellites around 1990.

Radio rebroadcasts which must, by their very nature, require a political agreement by the recipient country to use its territory for such purposes, are most economically accomplished by simply leasing telephone lines and the same number of telephone channels. The satellite channels normally have a bandwidth of 4 kHz for good quality voice, but by combining up to four such channels, music broadcast quality can easily be obtained, suitable for local rebroadcasting; a single channel is quite adequate for voice broadcasts. Satellite capacity may be on a full-time leased basis or on an hourly occasional use basis. A satellite service from one country to another can be provided by INTELSAT for US$65.00 per hour for a 4 kHz channel or US$110.00 per hour for a 16 kHz music channel. Additional receiving points can be added for US$28.75 and US$40.00, respectively.

If the broadcast company has a number of distant recipients, or for any reason prefers not to use the occasional use services using terrestrial networks, then other INTELSAT services are available that permit satellite services directly between the broadcasting premises having their own earth stations. For radio broadcasting alone such a system is inherently costly due to the cost of the earth stations, but there are cases where such a system is cost effective in which case the broadcaster could use the INTELSAT IBS or VISTA services described earlier.

6.2. INTELSAT LEASED TELEVISION SERVICES

INTELSAT's international leased television service came into being basically as a result of the occasional use service becoming congested and was therefore both beneficial to INTELSAT and the broadcasters. It was first introduced in 1981, but was expanded early in 1984 to accommodate a wider variety of technical applications. The major innovation had been the introduction of leased transponders for international television in 1981; the 1984 changes introduced a wider, but confusing, variety of bandwidths, transmission techniques, times, and priorities, which expanded the usefulness of the service.

INTELSAT now provides a variety of international video services, analog or digital, for full- or part-time lease, for distributing news and entertainment networks or video teleconferencing services. The charges for these vary widely depending on the transponder, satellite, and coverage. As a rule of thumb, each 72 MHz transponder costs US$1–2 million per year to signatories and within the United States at least, deregulatory policies are reducing the cost to broadcasters, as shown by the AT&T reduction of its annual charge by nearly 50% from around US$384,000 to US$204,000 per year. At the present time these services are predominantly between the more developed countries, but they can be expected to expand worldwide.

For domestic television each country may use any system it wishes but technical approval is needed from INTELSAT to obviate harmful interference problems. These national television programs leased from INTELSAT can also be legally received by others for a slightly higher charge. Services are generally designed to accommodate any requirements that are likely to be requested such as:

Global, hemispheric, zonal, or spot beams

Channels with 18, 24, 36, or 72 MHz bandwidth

Preemptible or nonpreemptible service

Long-term, short-term, or part-time leases

2, 5, or 7 year lease periods

Occasional use reception by nonparticipants.

The type of transponder determines the extent of geographic coverage, and can be global, hemispheric, zonal or spot beam depending on requirements and availability. In certain cases more connectivity can be arranged by using cross-strap beam facilities for east-to-east, west-to-west, or east-west connectivities if they are needed. Due to time zones, some networks require only one-way transmission at time-shared intervals to ensure full coverage of news between two countries.

Again, bandwidth and power requirements vary greatly. Some of the large networks are happy to have two larger earth stations in two countries exchanging programs, whereas other networks prefer distribution to a larger number of smaller earth stations. The former network distributes their programs by cable or radio, while the latter network can be received directly by much smaller earth stations similar to those for DBS or quasi-DBS. Generally speaking, the available power from the satellite can be concentrated into a smaller part of the earth's surface if a smaller beam is used. A small area of the earth is best suited to using a higher power spot beam so that the smallest and cheapest earth stations can be used so as to resemble the type of service envisaged by the DBS.

Connectivity arrangements also dictate the most suitable type of transponder. INTELSAT offers six types of connectivity; global, east-east, west-west, west-east, east-west, and certain combinations can be cross-strapped inside the satellite to provide connections between the Ku-band spot beams and the C-band hemispheric or zonal beams. The type of connectivity required relates to the type of video channel, of which there are three types: one-way only (simplex); two ways on a time-shared basis (half duplex); and full duplex which permits two-way transmission at the same time. Transponders with global, east-east, or west-west connectivity provide half-duplex video channels, while transponders with east-west or west-east connectivity can only provide simplex channels and full duplex service needs two simplex channels.

Transponder bandwidth, power, and earth station size determine the video signal quality, and there are technical as well as cost trade-offs between them. Options of bandwidth are usually 18, 20, 24, 36, or 72 MHz. The 72 MHz transponder can be used for transmitting three video carriers each of 24 MHz, but less power is available from the satellite and same frequency use can cause interference difficulties if these are not fully calculated and coordinated. Some typical television distribution networks are shown in Figure 6.2 and service growth as well as a list of present international television networks provided by INTELSAT is in Table 6.1.

Digital video over satellites has, until recently, been limited and largely experimental due to its high bandwidth requirements, but recent developments in bit rate reduction and coding techniques enabled INTELSAT to introduce a very high quality digital service in 1984. Initially bit rates of 68, 45, 34, 32, and 15 Mbps were offered, but lower bit rates are likely with the availability of more advanced codecs and it is likely that all digital video services will be using rates no higher than 30 Mbps within the decade.

The newer codecs provide fully acceptable quality video and do so using less transponder bandwidth as well as smaller earth station antennas than the present analog services. International quality standards for digital television have not yet been established, but subjective tests have established that a "fully acceptable quality" is better or equivalent to the current CCIR standards for FM analog transmissions.

These video capabilities are also consistent with the performance characteristics of the INTELSAT Business Services (IBS) of a 10 to -8 bit error rate between modems for 99% of the year. The overall quality is obtained from more error-correcting coding within the TV codes. Higher quality television distribution can be provided over present INTELSAT satellites with antennas as small as 3.5 m at Ku-band and 5 m at C-band. These digital video services are developing in a manner that is fully consistent with the Integrated Services Digital Network (ISDN) for which the world is presently aiming.

Digital video services can use a wide varity of earth stations such as the INTELSAT Standard A, B, F1, F2, or F3 at C-band, and Standard C, E1, E2, or E3 at Ku-band. Standard G earth stations can be used for reception only at either C- or Ku-band. As it is now possible for earth stations to

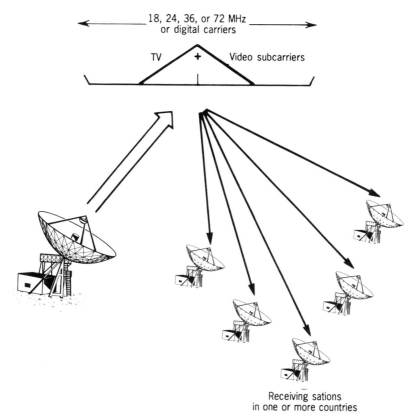

Figure 6.2. Typical international television networks.

access the satellites directly from the user premises, arrangements for this vary in each country depending on the policy of the country's government.

Some of the worldwide satellite television networks using INTELSAT are described shortly, and it is technically quite feasible for any person to receive these transmissions from the satellites provided that suitable receiving equipment is used. Many individuals are now doing this as a hobby, and there are many magazines and articles written on how it can be done. There are, however, a number of very difficult transmission parameters in use to impede the would-be recipient.

Down-link power from the satellite is the main parameter for determining the size of the earth station antenna. The higher the power, the smaller the antenna that can be used, and it was originally the aim of the DBS satellites to transmit sufficient power so that programs could be received at individual homes with antennas of about 60 cm in diameter—or a flat rooftop antenna. This size is decreasing as further technological advantages take place, and flat rooftop antennas as small as 25 cm are now being advertized.

TABLE 6.1 INTELSAT's International TV Leases

Service Growth

Year	Capacity Leased in MHz
1982	54
1983	162
1984	180
1985	270
1986	648
1987	786
1988	1,206

Principle Countries	Program	MHz	Beam	Satellite (°E)
Transpacific Television				
United States–Australia	TV7	18	Hemi	180
United States–Australia	TV9	24	Global	180
United States–Australia	TV10	24	Hemi	180
United States–Korea	AFRTS	36	Global	180
United States–Japan	JISO	18	Zone	180
United States–Japan	NHK	18	Zones	180
United States–Hong Kong	CNN	36	Hemi	180
Canada–New Zealand	NZBC	18	Global	180
Transatlantic television				
United States–United Kingdom	CBS	36	Crossed	332.5
United States–United Kingdom	ABC	36	Crossed	332.5

United States–United Kingdom	NBC	36	Crossed	332.5
United States–United Kingdom	VISNEWS	36	Crossed	332.5
United States–Germany		54	Crossed	325.5
United States–France	NETCOM	18	Global	332.5
United States–France	DTRE	18	Global	325.5
United States–France	USIA	72	Hemi	332.5
United States–Italy	EBU	24	Crossed	332.5
United States–Diego Garcia	AFRTS	36	Global	359
United States–United Kingdom/ /Panama	FBIS	36	Global	332.5
Spain–Argentina	TVE	72	Hemi	332.5

Intra American Television

United States–Brazil	USIA	72	Hemi	332.5

Intra European Television

United Kingdom–Finland	W. Smith	36	Spot	332.5
United Kingdom–Finland	Premier	36	Spot	332.5
United Kingdom–Finland	MTV	72	Spot	332.5
United Kingdom–Denmark	BBC	36	Spot	332.5
United Kingdom–Ireland	FSC	72	Spot	332.5
United Kingdom–Ireland	Scansat	72	Spot	332.5
France–Monaco	DGT	72	Hemi	332.5

TransAsia and Africa Television

France–Philippines	USIA	36	Global	66

Satellites also have different means of polarizing the transmitted signals, and the two common types are linear or circular polarization; different earth station feed systems are needed. The earth station receiver must be tuned to the correct frequency, have suitable demodulation or decoding equipment, sound receiving equipment, and lastly must use the correct television standard. The latter may well be the most difficult for the amateur enthusiast as different parts of the world use different standards, and conversion equipment is sometimes expensive. The three predominant standards are the National Television Standards Committee (NTSC), the phase alteration by line (PAL), and the sequence à memoire (SECAM) standards. The NTSC, using a black and white picture of 525 lines and 60 fields with a video bandwidth of 4.2 MHz, is the standard used by the United States, Japan, the Philippines, and some other countries. The PAL and SECAM standards both use a black and white picture of 625 lines by 50 fields, but have different color systems. PAL uses a color signal inverted 180° on alternate lines; SECAM sends the color information by separating the blue and red color data prior to transmission and transmitting it by two color subcarriers at 4.4 MHz and 4.25 MHz. Recombination is done on the television set. SECAM is used by most East European countries. PAL is probably the most common national standard now, but the NTSC standard is too far established to change at this stage of television development. It is to be hoped that when the future high definition TV (HDTV) comes into practice, that a single international standard will emerge. In addition to the various standards mentioned, a variety of coding systems are emerging, used by individual broadcasters they are mentioned later in the chapter.

The United States' Armed Forces Televison Network. The American Armed Forces Radio and Television Service (AFRTS) has been in existence for many years and in December 1982, AFRTS started transmitting its television service to the country's naval base on the remote Indian Ocean Island of Diego Garcia using the INTELSAT satellite located at 359°E. This location is just about the the only one that can "see" both the north eastern portion of the United States and the global beam's eastern edge at Diego Garcia.

A full 36 MHz global beam transponder was chosen, and several other bases in other countries later became recipients of the AFRTS program. This network is shown in Figure 6.3 which also shows a similar transponder lease in the Pacific ocean region that was established the following year, in 1983. AFRTS is now available almost worldwide; only countries between about 80° and 100°E are unable to receive it.

AFRTS includes programs of a number of United States news networks and one of the recipient countries, France, retransmits some of these over its TELCOM I satellite to other European countries and cable networks.

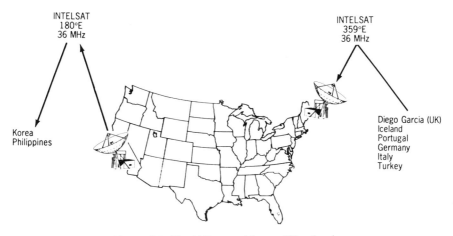

INTELSAT
180°E
36 MHz

INTELSAT
359°E
36 MHz

Korea
Philippines

Diego Garcia (UK)
Iceland
Portugal
Germany
Italy
Turkey

Figure 6.3. The U.S. armed forces TV network.

The American WORLDNET Television Network. WORLDNET is generated by the United States Information Agency (USIA) and was created to replace and improve the earlier information systems provided by films and information presentations usually sent out to their overseas agencies by air. The new television program is devised and provided by the Agency's headquarters in Washington D.C.

The service was inaugurated in April 1985 providing two hour programs to Europe using part-time video leased service through INTELSAT. In July 1986 the service was expanded to include programs in Spanish and Portuguese to Latin America. Both programs leave the United States through Standard A earth stations. The European program was originally received in Belgium and retransmitted round Europe through the EUTELSAT system. The European routing later changed and was sent to Europe using an existing United States–France video lease and then redistributed to Europe by a EUTELSAT satellite to American Embassies, Consulates, and Agencies which make WORLDNET available to cable networks, private homes, and video libraries all over Europe. The WORLDNET service is further extended to countries in the Indian ocean region using another INTELSAT lease between France and the Philippines. The Latin American program is received by Brazil's EMBRATEL Standard A earth station and retransmitted throughout Latin America. WORLDNET is free American programming; there are no royalties. An experienced anchorman from CBS News, the late Nelson Benton, was used to anchor an interactive news program, and WORLDNET is fast becoming a popular alternate television program, as well as being an excellent means of spreading United States government information beyond its borders. WORLDNET distribution is shown in Figure 6.4.

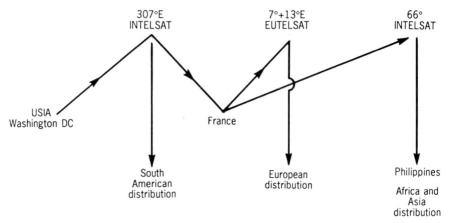

Figure 6.4. The U.S. WorldNet TV program.

Broadcast Companies. An increasing number of television broadcasting companies and cable networks are entering into international distribution of their national networks, especially between North America and Europe where large domestic distribution networks already exist. These include the American Broadcasting Company (ABC), Columbia Broadcasting System (CBS), National Broadcasting Company (NBC), Cable News Network (CNN), VISNEWS, and the French NETCOM, all of which have full-time television leases through INTELSAT. In the Pacific region, the number of television networks between the United States, the Far East, and Australasia is also increasing.

6.3. REGIONAL SATELLITE TELEVISION

The Americas. The basic geography and economic development of the United States and Canada put these two countries inherently ahead of any other market demand for television services. They already had extensive television networks before the arrival of satellites which merely enabled the networks to make an easy and economical leap with wider distribution.

Canada was the first to launch its own domestic satellite in 1972 and started distributing television programs, as well as other basic telecommunications services, to large areas of the nation that had previously been completely inaccessible to television. The United States satellites followed closely in 1973, and there now exists over North America over 100 television programs many of which cover all of North America, Hawaii, Alaska, Central American countries and the Caribbean. Many of these programs are redistributed by cable networks to offices, hotels, and private homes on a commercial basis and the distributors try to prevent pirating or free reception of their satellite signals by using coding or other techniques.

The present satellite systems transmitting domestic television services over the Americas are:

United States	AT&T—Telstar
	CONTEL/Amsat
	General Electric/RCA—Aurora and SATCOM
	General Motors/Hughes—Galaxy
	GTE—Spacenet and GStar
	PanAmsat
	Western Union—WESTAR (now General Motors/ Hughes)
Brazil	Brazilsat
Canada	ANIK
Mexico	Morelos
France	Telcom
	INTELSAT

There are other satellite systems planned, some firmly and others not so firmly, including:

Advanced Communications Corporation

American Pacific Broadcasting

National Exchange

Direct Broadcast Satellite Corporation

United States Satellite Broadcasting Corporation

Tempo Enterprises

Rainbow (RSI)

Satellite Syndicated Systems (SSS)

National Christian Network

Spotnet

Dominion Video Satellite

Some of these have been approved by the FCC, some are not likely to materialize, and all were forcibly provided with extra planning time by the lack of launch facilities following the shuttle disaster early in 1986.

As well as the television networks, there are a number of satellite news gathering operators whose business is to gather news events for the larger networks. This new line of business uses satellite facilities that are available mainly on an occasional use or shared basis.

Europe. Throughout Europe, governments have seen the potential for the new television industry to create jobs and expand local industries—even though it threatens cultural sovereignty and national languages. Most countries would like to retain control over their television networks, but they now generally recognise the practical problems of doing so. The present trend towards deregulation and privatization of European industries is also helping towards a much wider and less controlled number of television networks within Europe which cross national frontiers. The general attitude seemed first to delay deregulation as much as possible while local industries built up to counteract the U.S. onslaught of equipment and programming. This has now been replaced by mainly open market policies that are permitting a proliferation of new television programs within Europe, both European and American in origin. The expansion of television programs in Europe has been much faster than had been expected.

Programming probably evoked the strongest feelings as U.S. programs were already available whereas Europeans had very little of their own until recently, and certainly insufficient to fill up 30 channel cable television networks. In the United Kingdom, there has been increasing demand over many years to provide more television channels, and its common language with North America made it a prime target for U.S. programming. France restricts foreign programming to a maximum of 33% on any cable network, and a program channel can only be imported if a French program channel is exported in exchange. Television on the French TELECOM satellites are also received in French-speaking countries overseas such as Haiti, Madagascar, and Mauritius as well as its own departments of Martinique, Guadaloupe, French Guiana, and Reunion.

European nations are still debating the relative merits of cable and DBS technologies as well as a number of different transmission standards. Cable networks seemed most popular at one time as they permitted governments to have closer control over the networks, but again, deregulation trends have opened the entire market. Satellite distribution of television within Europe has been growing rapidly over the last five years and had obtained a very large audience even before the arrival of the first DBS satellite (Germany's TVSAT, in 1987 which was not successful due to component failure). The debate over the merits of DBS over cable and other satellites continues, but the investment in DBS has been very large and mainly government sponsored.

The first satellite television distribution networks used INTELSAT spot beams or the Eutelsat satellites. Luxembourg's ASTRA satellite and the British Satellite Broadcasting satellite will shortly be adding many more transponders for television networks in Europe. In addition, Eurosatellites' DBS satellites being made for Germany, France, and Sweden will provide yet more capacity.

The costs of television distribution have largely been set the the INTELSAT charges of around US$1–2 million per year, and in late 1987 INTEL-

TABLE 6.2 Satellite Television in Europe

Originator	Program

1. *EUTELSAT I, F-1 at 13°E*

Originator	Program
Belgium	Filmnet
France	TV5
Germany	3Sat
Germany	Sat1
Italy	RAI 1
Luxembourg	RTL Plus
Switzerland	Teleclub
United Kingdom	Sky Channel
United Kingdom	Super Channel
United States	Worldnet

2. *EUTELSAT I, F-2 at 7°E*

Originator	Program
United States	Worldnet
EBU	Eurovision

3. *EUTELSAT I, F-4 at 10°E*

Originator	Program
Norway	NRK
Italy	RAI
Spain	TVE1

4. *France's TELECOM I at 355°E*

Originator	Program
France	Canal J
France	La Cinq
France	M6

5. *INTELSAT VA F-11 at 332.5°E*

Originator	Program
France	DGT
Spain	Canal 10
United Kingdom	Arts Channel
United Kingdom	Anglovision
United Kingdom	BBC1/2
United Kingdom	Childrens Channel
United Kingdom	Lifestyle
United Kingdom	Screensport
United Kingdom	Radio Nova
United Kingdom	Premiere
United Kingdom	TV3-Scansat
United Kingdom	W.H. Smith
United Kingdom	MTV

6. *INTELSAT VA F-12 at 60°E*

Originator	Program
Germany	3Sat
Germany	Eins Plus
Germany	Eureka
Germany	Music Box
Germany	WDR3
Germany	BR3

TABLE 6.2 (*Continued*)

Originator	Program
7. INTELSAT V, F-2 at 359°E	
Norway	SVT-1
Norway	SVT-2
Norway	New World
Israel	
8. INTELSAT V, F-4 at 325.5°E	
Spain	2 Channels
9. USSR's GORIZONTs at 349° and 356°E	
U.S.S.R.	Moskva TV

SAT was providing well over half of Europe's satellite television services as can be seen in Table 6.2, although it should be noted that both satellite and transponder changes take place quite frequently.

Some coverage diagrams for the EUTELSAT, INTELSAT, and TELCOM satellites are shown in Figure 6.5. Luxembourg's ASTRA and the British Satellite Broadcasting satellites are very similar except that beam coverages vary slightly. The planned television satellites which could produce a glut of capacity over the next few years are:

Country	Satellite	Year	Channels	Power (dBW)
Luxembourg	ASTRA 1	1988	16	50
France	TDF 1	1988	4	60–65
Germany	DFS	1989	4	60–65
Sweden	Tele-X	1988	4	60–65
United Kingdom	BSB	1989	16	64
EUTELSAT	II A&B	1990		45–51
Ireland	Atlantic	1991		63–65

Asia-Pacific. Regional braodcasting activities in the countries of Asia and the Pacific between the Gulf of Suez and Hawaii have been coordinated since 1964 by the Asia-Pacific Broadcasting Union (ABU) which was set up to assist the development of radio and television in the region. Originally one country could only have one membership in the ABU which of course restricted broadcasting companies in those countries where more than one such company existed. In 1983 this changed and each country is now permitted two memberships, and there are presently 70 members from 47 countries and territories.

The ABU has greatly influenced broadcast programs and technical standards which has been particularly useful. Its news program *Asiavision* takes contributions from among its members, thus enabling countries to see events in their neighbor's countries as well as their own. Vision only is transmitted allowing each country to provide its own sound to prevent the unwanted ingress of political propaganda. To the present time satellite links have been almost entirely through INTELSAT.

ARABSAT provides some television programs to its own members and the Indonesian PALAPA satellite provides some services within its area of

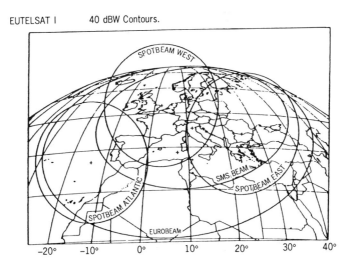

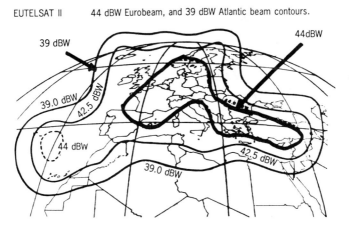

Figure 6.5. Satellite coverages of Europe.

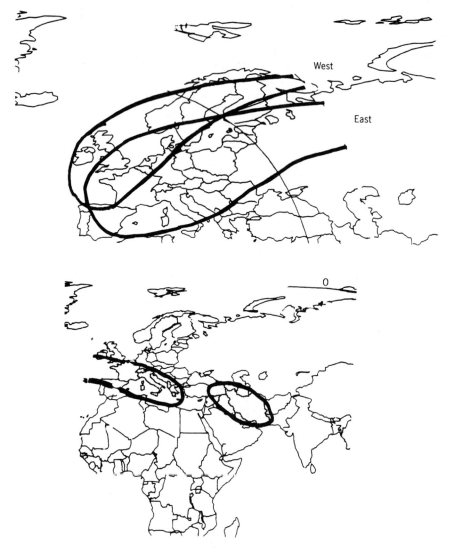

Figure 6.5. (*Continued*)

coverage. There is thus considerable scope for the expansion of television broadcasting in this vast area of the world. There are now some potential satellite service suppliers other than INTELSAT being planned, and the most likely possibility could be the AsiaSat satellite mentioned in Chapter 3, but at this stage insufficient is known of its coverages and services to determine whether full regional coverage will be feasible, or at what cost.

Other Parts of the World. Many countries already have one national television channel which is usually distributed over wide areas by means of INTELSAT domestic leases, or through individual satellite systems in

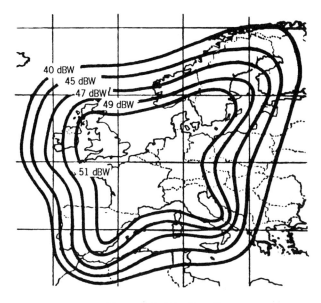

Figure 6.5. (*Continued*)

Indonesia, India, Mexico, Brazil, and Australia. The Arab regional satellite system, ARABSAT, is also available for distributing television within its own region, but on the whole the remainder of the world is wide open for the spread of television programs.

Television program distribution is of great interest to developing countries for education and general information purposes. Satellite distribution can, or could be enormously useful in this process. Existing facilities, which are mostly general in purpose, are very cost-effective. One important aspect at the present time is that there are no international agreements on transmission standards for any of the programs available. If popularity is a guide, there is growing support for a system known as multiplexed analogue component (MAC) which was originally developed by Britain's Independent Broadcasting Authority, and approved by the European Broadcasting Union (EBU). MAC basically divides a single frequency into different time slots into which are inserted codes for picture chrominance, luminance, and sound. This breaks with conventional techniques which use different frequencies for the different elements. However, there are several versions of MAC and further agreements are still needed.

One major potential supplier of international television is the Soviet Union whose INTERSPUTNIK satellites have been relaying Soviet and member countries' television for many years. Their new TOR satellites are expected within this decade and the sheer power superiority of these over INTELSAT transponders for television purposes could make INTERSPUTNIK a major competitor in the area of international television distribution.

7

MOBILE APPLICATIONS

Mobile satellite services (MSS) have commonly come to mean the use of satellites with cars and road vehicles to extend what is called "cellular" radio in the United States, whereas they also include maritime mobile satellite services (MMSS) for ships at sea, and aeronautical mobile satellite services (AMSS) for aircraft. These also include radiodetermination and radionavigation services. Mobile satellite services to land vehicles are termed land mobile satellite services (LMSS). All these services have been allocated bands of radio frequencies which are summarized in Table 7.1 as a guide, but these frequencies contain restrictions in certain countries and regions. These frequency bands, as indeed are all radio frequency bands, are organized by the International Frequency Registration Board (IFRB) under the auspices of the ITU, and the World Administrative Radio Conference (WARC).

Some countries that have their own satellites have been developing their mobile capabilities to suit their own requirements and incorporating mobile service capabilities in their satellites. International mobile satellite services are provided, up to this time, only by INMARSAT. Although mobile applications are still strictly limited due to satellite availability, these will be expanding quickly within the next two years.

7.1. MARITIME MOBILE APPLICATIONS

Satellite services for ships at sea started in 1976 using MARISAT satellites launched and owned by COMSAT General Corporation of the United States in conjunction with some other U.S. telecommunication companies (RCA, Western Union, and ITT). Global coverage of all oceans was achieved in 1979 using three satellites, one being located centrally over each of the three major oceans, the Atlantic, Indian, and Pacific; these are still operational and in good working order over a decade later. An international organization (INMARSAT, see also Chapter 2) was also formed in 1976

TABLE 7.1 Mobile Satellite Services Frequency Bands

Frequency Band	Type of Service
	MHz
235 –322	Aeronautical
335.4–399.5	Aeronautical
339.5–400.5	Radionavigation
406.0–406.1	Beacons, emergency locators up-link
608 –614	Distress and safety up-link
806 –890	Aeronautical
942 –960	Aeronautical
	GHz
1.215– 1.240	Radionavigation down-link
1.240– 1.260	Radionavigation down-link
1.530– 1.560	Beacons, emergency locators down-link
1.535– 1.543	Maritime down-link
1.544– 1.545	Distress and safety down-link
1.559– 1.626	Radiodetermination
1.636– 1.645	Maritime up-link
1.542– 1.558	Aeronautical down-link
1.610– 1.626	Aeronautical
1.631– 1.634	Beacons, emergency locators up-link
1.636– 1.644	Maritime up-link
1.644– 1.660	Aeronautical up-link
1.645– 1.646	Distress up-link
2.500– 2.535	Aeronautical down-link
2.655– 2.690	Aeronautical up-link
5.000– 5.250	Aeronautical
7.250– 7.375	Aeronautical down-link
7.900– 8.025	Aeronautical up-link
14.000–14.500	Aeronautical up-link
15.400–15.700	Aeronautical up-link
19.700–21.200	Distress, safety and emergency
29.500–30.000	Distress and safety
30.000–31.000	Beacons, emergency locators
39.500–40.500	Beacons, emergency locators
50.400–51.400	Distress and safety
66.000–71.000	Beacons, emergency locators

which started operations by leasing facilities on MARISAT, INTELSAT, and the European Space Agency's MARECS satellites until 1989 when INMARSAT's own satellites are expected to become operational. The mid-1988 satellite status is shown in Table 2.3.

As mentioned earlier in Chapter 2, INMARSAT's service is essentially a number of voice/data channels for use by ships at sea either to communicate to land, or between ships at sea. Maritime uses of a single voice/data communications channel include:

Emergencies and SOS calls
Estimated time of arrivals
Navigation data
Daily reports from ships
Weather data
Cargo stowing plans
Maintenance and spare parts
Private telephone calls
Oil drilling data
Engine and fuel monitoring data
Seismic data
Offshore early warning data
Slow-scan television
Fleet management
Shipping agency coordination
Ship routing
Port instructions
Medical advice
Crew rosters
News bulletins
Teletext
Travel agency communications

Communication with ships at sea has a very brief history considering that all of this has taken place during the twentieth century since the first transatlantic radio message was transmitted by Marconi in 1901. Even as late as the mid-1950s, only a few ships were equipped with a radio telephone and these were mainly the large ocean liners. Apart from these ships, the sole means of communication with ships was the Morse key; not all ships had the ability to transmit even this! The foregoing list of new uses, therefore, is an enormous advance over what was possible prior to 1976 when satellite services first became available.

As a guide to the ways in which the ships are now using INMARSAT's satellite services, the types and number of ships with ship satellite terminals are shown in Table 7.2 which is a summary of data supplied by INMARSAT (Ocean Voice, April 1988). The number of ship terminals registered by national flags are shown in Table 7.3.

To describe many of the maritime applications for satellite communications would be like describing at length the ways in which one can use a telephone. The main event resulting from satellite communications is that a satellite makes possible what was hitherto impossible—a good, reliable

TABLE 7.2   Maritime Ship Terminals

Annual Growth in Total of Ship Terminals

Year	Total	Year	Total
1976	34	1982	1,509
1977	92	1983	2,124
1978	166	1984	3,004
1979	293	1985	3,965
1980	502	1986	5,051
1981	905	1987	6,331
		1988 (March)	6,705

	Number of Terminals	
Type of Ship	1987 (March)	1988 (March)
Bulk	787	903
Container	399	466
General cargo	773	980
Tanker	1,260	1,502
Liquid gas carrier	149	184
Oil rigs	271	297
Barges	126	149
Fishing vessels	261	494
Seismic survey	131	117
Research	81	109
Passenger	121	138
Yachts	350	467
Government ships	166	218
Land terminals	329	531
Tugs	64	81
Miscellaneous	52	69
Total	5,320	6,705

TABLE 7.3   Registered National Flag Ship Terminals (March 1988)

Country[a]	Total[a]	Country	Total
Algeria	1	Canada	66
Angola	1	Central African Rep.	1
Antigua	5	Chile	3
Argentina	12	China	39
Australia	57	Colombia	1
Austria	6	Comoros	2
Bahamas	154	Congo	1
Bahrain	10	Cyprus	82
Belgium	31	Denmark	79
Brazil	29	Ecuador	4
Brunei	10	Egypt	13
Bulgaria	40	Equatorial Guinea	1
Burma	3	Finland	28
Burundi	1	France	117

TABLE 7.3 (*Continued*)

Country[a]	Total[a]	Country	Total
Gabon	3	Norway	194
Germany (DR)	13	Oman	21
Germany (FR)	146	Panama	778
Greece	180	Papua New Guinea	4
Guinea	2	Peru	1
Honduras	4	Philippines	120
Iceland	4	Poland	48
India	42	Portugal	11
Indonesia	2	Qatar	18
Iran	11	St. Vincent	4
Iraq	5	Sao Tome and Principe	1
Ireland	3	Saudi Arabia	94
Israel	1	Senegal	2
Italy	104	Singapore	117
Ivory Coast	3	Solomon Islands	1
Japan	672	Somali Republic	1
Jordan	10	South Africa	1
Korea (Republic of)	34	Spain	91
Kuwait	24	Sri Lanka	8
Lebanon	19	Sweden	28
Liberia	668	Switzerland	7
Libya	5	Taiwan, Prov China	76
Madagascar	5	Thailand	1
Malaysia	9	Togo	3
Mali	1	Turkey	9
Malta	5	United States	1032
Mauritania	4	U.S.S.R.	280
Mauritius	1	United Kingdom	680
Mexico	23	United Arab Emerates	40
Monaco	3	Vanuatu	16
Morocco	13	Venezuela	18
Netherlands	130	Yemen Arab Rep.	2
New Zealand	11	Yugoslavia	27
Nigeria	2	Zaire	12

[a] Total terminals = 6,705; total countries = 96.

means of communication with ships, and now aircraft. Now that the facility is available, maritime applications are mainly a matter of system growth and some standardization for such things as safety at sea, and, of course, the national interest of individual countries. Maritime satellite communications have brought to ships what has been available on land for many years, with the additional possibilities for full computer links to shore.

The most important application of maritime satellite communications is that of safety at sea. Prior to maritime satellites, safety at sea relied almost entirely on a ship's radio officer sitting down at a Morse key and tapping out the famous SOS (. . .−−−. . .) signal together with the ship's name and estimated position for as long as it remained possible to do so, or until the signal was acknowledged and help was on its way. This system saved many thousands of lives but the International Maritime Organization's (IMCO) future global maritime distress and safety system (FGMDSS), starting on 1 July 1991 will do much, much more. One of the most important features will be that ships will no longer need to maintain radio watches except in certain maritime regions where ships must receive maritime safety information affecting the area in which they are sailing.

INMARSAT's new enhanced group call (EGC) service will provide a reliable means of sending messages either to groups of ships, or to specified geographic areas via a high-capacity and reliable satellite broadcast channel. The EGC's SafetyNet service will send maritime safety information, as needed, for routine weather forecasts, storm warnings, and information on any distress alerts. For example, a hurricane warning could be sent to a specified area in the Caribbean covering the region of its anticipated path. This information could be updated every few hours and readdressed to cover the latest path predictions. Only ships within the area specified by the meteorological office would receive the transmitted message. This would filter out much useless information from an individual ship's safety procedures. Small ships will also be affected by these procedures, and are required to be equipped before 1997.

Commercially, INMARSAT's Fleetnet service provides EGC calls to be used by shipping companies, or private subscribers, to send information to ships such as news, market services, market prices, and information updates to selected ships on a selectable broadcast basis.

Similar to the situation concerning private international competition to INTELSAT, there are some private entrepreneurs who see a viable future in providing mobile services, including services to ships and small craft. One of these, the Geostar Corporation of the United States founded in 1982, plans to provide a variety of services including

Navigational positioning
Radiolocation
Emergency location
Warning of approaching land
Warning of potential collision
Aircraft approach guidance
Two-way digital message service

These include the whole scope of mobile applications; maritime, aeronautical and land mobile.

Geostar had hoped to start operations in 1986 using a relay package developed by RCA Astro-Electronics and carried on GTE Spacenet's GSTAR 2 satellite, but the payload failed and replacement payloads are planned. Although not operational at the time of writing, Geostar is economically viable and could become an international provider of such services, especially in the vicinity of the United States.

7.2. AERONAUTICAL MOBILE APPLICATIONS

Like most large industrial applications, aeronautical telecommunications emerged most rapidly in the United States. Aeronautical Radio Inc. was formed in 1929 by the government and airline industry to operate aeronautical telecommunications within the United States. It is limited to a maximum of 20% foreign ownership which makes it unsuitable for international cooperative ventures such as required for international aeronautical services. An international organization was formed in 1949 by 11 international airlines called the Société internationale des telecommunications aeronautiques (SITA) whose aims are to ensure rapid transmission of information relating to airline operation and control. SITA operates telecommunications networks linking over 25,000 teletype or computer terminals in some 16,000 airline offices in 1000 cities. There are now 258 member airlines in 160 countries with two data processing centers, one in London and one in Atlanta, Georgia.

When international satellite services became a possibility in the early 1960s, a subsidiary of Aeronautical Radio Inc. (ARINC) was created in the United States called Aviation Satellite Corporation (AVSAT) which is able to have much larger foreign investment than its parent organization. AVSAT was formed to create international satellite services for the airline industry that were needed for air traffic control, airline operations, airline administration, and also private telephone for airline passengers on an aircraft. Until early 1988, almost all telecommunications to aircraft are obtained using the UHF band directly between the ground and the aircraft.

The need for aeronautical satellite services has been seen for many years and radio frequency spectrum has been allocated. Due to the delays in starting services, the frequencies allocated were not used, and pressures from other potential users of the spectrum caused the 1987 WARC to reduce the allocated frequency band, and necessitated sharing of about one third of the allocated band with the land mobile satellite service.

Satellite communication for aircraft in the sky is just about the saddest story in the applications of satellite communications. It started as a preliminary study in 1964 by some interested U.S. corporations and nearly a quarter of a century later in 1988 services are just beginning to emerge. In 1964, a cooperative industry study started in the United States consisting of ARINC, its international subsidiary AVSAT, Pan American Airways, Hughes Aircraft, Bendix, Boeing and COMSAT, the new U.S. satellite operating corporation. Over 23 years later in 1988, it seems that satellite communications with aircraft will finally become a reality.

As far back as 1964 the study showed the feasibility of air-to-ground voice and data by satellite, and even recommended that the Early Bird III (later known as INTELSAT III) be modified to provide this service. The list of U.S. Government agencies, public corporations, and international agencies that became involved in the subsequent delay is exhaustive and includes the Department of Transportation, the FCC, the Federal Aviation Agency (FAA), ARINC, COMSAT, the Office of Telecommunications Policy (OTP), the Inter-Agency Group on International Aviation, the International Civil Aviation Organization (ICAO), the U.S. Department of State, airlines and communications companies, and the House Appropriations Subcommittee. These deliberations yielded some noteworthy nonevents including:

1964 The first study of the practicality of communicating to and from aircraft via satellites was made by a section of U.S. industry (COMSAT, Bendix, Hughes Aircraft, and PanAm).

1966 Successful tests of VHF communications took place using the NASA ATS-1 satellite Robert R. Bohannon, *An International Airline's View of Satellite Techniques*, Paper at IEEE EASCON 1974).

1968 An INTELSAT system engineering study by COMSAT in which satellite bids were received, evaluated, and design feasibility confirmed, but INTELSAT's ICSC did not pursue it.

1968 The International Civil Aviation Organization (ICAO) formed the panel on the Application of Space Technologies Relating to Aviation (ASTRA).

1970 The United States and some European countries discussed the establishment of an Atlantic air navigation system to be called AEROSAT, and agreed on the use of UHF technology. A Memorandum of Understanding (MOU) was signed on behalf of the United States by NASA and the FAA and the European Space Research Organisation (ESRO) signed on behalf of Europe. Meanwhile in the United States, ARINC and COMSAT had proposed a VHF satellite system. "When the time came for the White House to sign the Agreement, it argued that it would be unfair for the Government to be the owner of a commercial venture, since it could ask industry to do it and then lease the industry services" (Kildow 1973: 80).

1970–1973 Political and technical problems resulted in AEROSAT funds being denied by the U.S. Government.

1973 The United States, Canada, and most European countries agreed on a further approach for an aerosatellite program called AEROSAT (AEROSAT Memorandum of Understanding, Annexe 1).

1976 An AEROSAT RFP was released (AEROSAT Satellite Specifications RFP SPO-001).

1977 Congress eliminated funding for the AEROSAT program and authorized US$1 million for another feasibility study.

1986 INMARSAT announced plans to start an aeronautical service by 1988 with flight tests starting in 1987.

1987 The FCC ordered its INMARSAT signatory (COMSAT) not to participate in these trials, but trials were permitted by other U.S. companies such as ARINC.

Bearing in mind that an aeronautical satellite system has been found feasible on numerous occasions, and also that in many areas of the world it is virtually impossible for aircraft to communicate with the ground near or underneath them, these delays seem incomprehensible. I believe that this delay was mainly due to arguments within the United States concerning ownership and operating functions of the proposed services. COMSAT, as the U.S. signatory to both INTELSAT and INMARSAT, had no desire to see another international satellite organization, but the airline industry wished to maintain some independence from what the airlines consider monopolistic companies and foreign Post and Telecommunications administrations. The airlines seemed unwilling, however, to bear the cost of such a system on their own. Hopefully, these problems are now resolved sufficiently to permit the long-awaited services to commence.

It is good to relate that in mid-1988, trials have at last started for services between some larger aircraft and the ground on some of the major transoceanic routes; unfortunately they will still not be available on many of the routes where there are no communications at present.

Equipment characteristics have been developing for aircraft despite numerous technical constraints. One of the severest concerns the antennas on the aircraft which must be small, and fit on the outside of the aircraft with no structural changes being made to the aircraft, and penetration of the aircraft skin must be very severely restricted because of structural and pressurization problems. Several aircraft antenna systems are now available but the one being installed by Boeing on its new 747-400 airliners consists of three antennas with one on the top of the aircraft and one on each side forward of the wings. The side-mounted antennas are flush-mounted panels which are scanned phased arrays, while the one on top is a keyhole antenna looking fore and aft to ensure all round coverage. INMARSAT and international manufacturers are cooperating to achieve agreement on standards for the equipment (Donald K. Dement, *Developing a Global Aeronautical Satellite System*, Paper presented at Mobile Satellite Conference 1988, NASA Jet Propulsion Laboratory).

INMARSAT's aeronautical services consist of two types; the first basic service provides low speed data links for airline operations, weather, air traffic control, position and aircraft performance monitoring and, of course, safety. The second is somewhat more complex and provides higher speed data and telephone communications including services for passengers. Airlines are already ordering facilities on new aircraft, and the SITA plans to build at least six earth stations to provide global coverage for its members. Peter Wood and Keith Smith, *World-Wide Aeronautical Satellite Communications*, Paper presented at Mobile Satellite Conference 1988, NASA Jet Propulsion Laboratory).

7.3. LAND MOBILE APPLICATIONS

There are, as yet, no commercially available satellite services for mobile services for vehicles although they have been under consideration for many years. This has been largely due to the technical problems involved, and the expense of developing earth stations that are sufficiently small, powerful, and also have reasonable flexibility with regard to pointing towards the satellite. Earth stations for maritime and aeronautical uses are leading the way, but whether commercial satellite ventures towards mobile earth stations and services will develop within this decade is doubtful; if they do, it will almost certainly be on a trial basis using existing satellites such as INMARSAT's. Tests performed late in 1987 in the United States provided about 75% successful voice connections with moving trucks. The satellite-to-mobile success rate was 92%, but the truck-to-satellite link was 75%. The other 25% suffered interruptions when the trucks passed trees and buildings that intervened between the truck and the satellites. (David C. Nicholas, *Land Mobile Satellite Propagation Results*, Paper presented at Mobile Satellite Conference 1988, NASA Jet Propulsion Laboratory.)

There are mobile telecommunications services in use, especially in the United States and Europe, usually referred to as cellular radio, which permit a single mobile terminal to access distant hubs. National as well as international agreements on frequencies are needed before satellite mobile applications are likely to be seen. Many feasibilty and market research studies have been performed in North America and Europe, which foresee a good market for such services at some time in the future, and some private companies have been formed with the view to operating such satellites on a national basis when it becomes possible to do so. In the United States, MSS applications filed with the FCC include those from:

Global Land Mobile Satellite Inc.
Globesat Express
Hughes Communications Mobile Satellite Inc.
Mobile Satellite Corporation
Mobile Satellite Service Corporation
Omninet Corporation
Skylink Corporation
Wismer & Becker/Transit Communications Inc.

Early in 1987, the FCC in an attempt to end a long impasse, asked all these to form a consortium which would be authorized 27 MHz of radio frequency spectrum in the L-band shared with the aeronautical service. Consortium members would be required to finance the estimated US$400 million for the system. A seven-company consortium, the American Mobile Satellite Consortium, was formed in May 1988 consisting of:

Hughes Communications
MCCA
McCaw Space Technologies
Mobile Satellite Corporation
Satellite Mobile Telephone
Skylink Corp
Transit Communications

This consortium hopes to start offering interim satellite communications as early as 1989 using existing satellites until its dedicated satellite is available around 1993.

Of these, Omninet started a radiodetermination service in May of 1988 using GTE's GStar satellite for relaying alphanumeric messages between trucks and their home bases.

In Europe, several countires have performed research and development into various applications for mobile satellite applications both on the satellites and on ground communications equipment. The European Space Agency's PROSAT project coordinates these activities, and numerous companies are developing equipment to work with PROSAT when it is completed. These mobile networks are emerging as a follow-on to national and European cellular radio networks.

8

NATIONAL
TELECOMMUNICATIONS
VIA SATELLITE

Although telecommunications were first used for international services, mainly to cross oceans, it was quickly seen that satellites could equally well be used to improve telecommunications on a national level in many countries. In fact, this application had already been foreseen by the founding members of INTELSAT when they were formulating the INTELSAT agreements between 1964 and 1969. Apart from national telecommunications within the Soviet Union, it was the INTELSAT system that was first used by individual countries for national development applications. The growth of the use of satellites for national telecommunications services is shown in Table 8.1.

In 1980, INTELSAT conducted a short study of the number of countries that were potential users of satellites for domestic purposes and also made a forecast of how INTELSAT's domestic services might grow over the next eight years. A far from comprehensive list of potential countries was made and is shown in Table 8.2. The conclusion of this brief, and deliberately simple study, was that very many countries could use satellites for domestic purposes, and the market had hardly started to be tapped. When it came to transferring this information into a forecast, again a simple approach was taken and is shown in Figure 8.1 but with the end 1987 actuals added to it. It is interesting to note that two things are quite apparent:

1. Countries that use satellites are using more capacity than was forecast
2. There are far less countries using satellites than was forecast; the market openings remain

INTELSAT now expects about 50 countries will be using 150 transponders by 1993 which is also shown graphically in Figure 8.1. This anticipates further high growth.

TABLE 8.1   Growth in the Use of Domestic Satellite Services

INTELSAT Domestic Service Growth

Year	Countries	Units of 36 MHz		
		Leased	Sold	Total
1974	1	1.00		1.00
1975	3	3.00		3.00
1976	7	6.50		6.50
1977	12	10.25		10.25
1978	14	13.75		13.75
1979	15	15.50		15.50
1980	14	17.50		17.50
1981	18	20.00		20.00
1982	23	36.50		36.50
1983	23	37.75		37.75
1984	28	48.50		48.50
1985	27	42.50		42.50
1986	29	28.25	36	64.25
1987	33	27.00	50	77.00
1988	37	43.00	82	125.00

37 Countries Using INTELSAT—1988

Country	MHz	Country	MHz	Country	MHz
Algeria	36	Iran	360	Pakistan	9
Argentina	360	Israel	216	Peru	252
Australia	20	Italy	288	Portugal	72
Bolivia	72	Ivory Coast	9	South Africa	72
Chile	72	Japan	72	Spain	66
China	315	Libya	36	Sudan	36
Colombia	45	Malaysia	54	Thailand	72
Denmark	36	Morocco	36	Turkey	144
Ethiopia	72	Mozambique	9	United Kingdom	72
France	119	Niger	72	United Nations	9
Gabon	108	Nigeria	108	Venezuela	72
Germany	432	New Zealand	18	Zaire	36
India	171	Norway	468		

Countries with National Telecommunications Satellites

1965	U.S.S.R.
1974	Canada
1975	United States
	Indonesia
1982	India
1984	Japan
1985	France
	Brazil
	Australia
	Mexico
1988	Luxembourg

TABLE 8.2   Potential Countries for Domestic Satellite Services

Country	Area (sq. miles)	Basic Terrain	Population Millions
Afghanistan	250	Mountain	21
Angola	481	Varied	7.2
Bangladesh	55	Varied	34.2
Benin	43	Forest	3
Botswana	232	Desert	0.7
Brunei	2	Forest	0.5
Burma	262	Forest	31.8
Burundi	11	Forest	4
Central African Rep.	241	Varied	2
Congo	122	Forest	1.5
El Salvador	8	Varied	4
Ethiopia[a]	472	Varied	30
Gabon[a]	103	Varied	10
Ghana	92	Forest	11
Guatamala	42	Varied	6.5
Guyana	83	Forest	1
Honduras	43	Varied	3
Iran[a]	636	Varied	34
Iraq[a]	167	Varied	12
Ivory Coast[a]	125	Forest	7
Kenya	225	Varied	14
Korea (R.O.)	38	Varied	36
Lesotho	11	Mountain	2
Liberia	43	Forest	2
Madagascar	227	Varied	3.5
Malawi	46	Varied	5.4
Mali	479	Desert	6
Mozambique[a]	303	Varied	10
Napal	54	Mountain	14
New Zealand[a]	104	Varied	3.2
Papua New Guinea	178	Forest	3
Paraguay	157	Varied	3
Philippines[a]	115	Islands	14.5
Senegal	76	Varied	5.3
Sierra Leone	28	Forest	3.2
South Africa[a]	472	Varied	3.5?
Somalia	246	Varied	3.4
Sri Lanka	25	Forest	14
Tanzania	365	Varied	16
Togo	22	Forest	2.5
Tunisia	64	Desert	6
Turkey[a]	301	Varied	41.6
Upper Volta	106	Desert	6.4
Uruguay	69	Varied	2
Vietnam	123	Varied	48
Zambia	290	Varied	3.5
Zimbabwe	150	Varied	6.8

[a]Indicates countries that now use INTELSAT for domestic use.

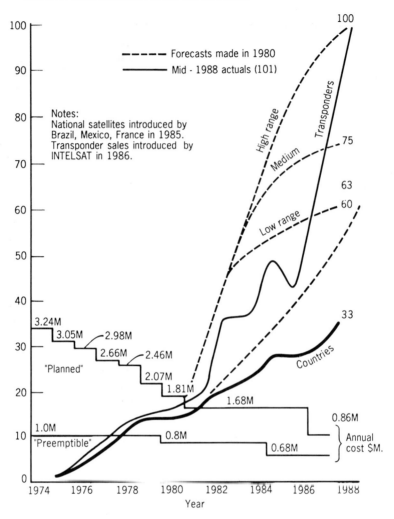

Figure 8.1. INTELSAT's domestic service growth.

TABLE 8.3   Domestic Satellite Systems in 1988

Country	Bandwidth Available (MHz)	Satellite System Used	First Started
Algeria	36	INTELSAT	1975
Argentina	180	INTELSAT	1982
Australia	675	AUSSAT	1980[a]
Bolivia	72	INTELSAT	1988
Brazil	864	BRAZILSAT	1976[a]
Central African Rep.	9	INTELSAT	1986
Chile	72	INTELSAT	1977
China	315	INTELSAT	1985
Colombia	45	INTELSAT	1978

TABLE 8.3   *(Continued)*

Country	Bandwidth Available (MHz)	Satellite System Used	First Started
Denmark	18	INTELSAT	1977
Ethiopia	72	INTELSAT	1988
France	117	INTELSAT	1975
France	536	TELECOM	1985
Gabon	108	INTELSAT	1986
Germany	432	INTELSAT	1986
India	144	INTELSAT	1982
India	432	INSAT	1984
Indonesia	864	PALAPA	1976
Iran	216	INTELSAT	1986
Israel	216	INTELSAT	1986
Italy	216	INTELSAT	1986
Ivory Coast	9	INTELSAT	1987
Japan	72	INTELSAT	1987
Japan	7860	CS-2	1977
Japan	50	BS-2	1978
Libya	36	INTELSAT	1982
Malaysia	54	INTELSAT	1975
Malaysia	36	PALAPA	1979
Mexico	1584	MORELOS	1980[a]
Morocco	36	INTELSAT	1982
Mozambique	9	INTELSAT	1986
New Zealand	18	INTELSAT	1986
Niger	72	INTELSAT	1986
Nigeria	108	INTELSAT	1975
Norway	252	INTELSAT	1975
Oman	36	ARABSAT	1978
Peru	198	INTELSAT	1978
Portugal	72	INTELSAT	1982
Saudi Arabia	180	ARABSAT	1985
South Africa	72	INTELSAT	1986
Spain	36	INTELSAT	1978
Sudan	36	INTELSAT	1977
Thailand	72	INTELSAT	1982
Thailand	36	PALAPA	1978
Turkey	144	INTELSAT	1986
United Kingdom	225	INTELSAT	1984
United States	2592	AT&T, Telstar	1983
United States	2592	Comstar	1976
United States	2520	AMSAT	1985
United States	3456	GE/RCA	1975
United States	2592	GM/Hughes	1983
United States	6294	GTE Spacenet	1984
United States	1720	MCI/SBS	1980
United States	1296	PanAmSat	1988
United States	2592	WESTAR	1974
United Nations	9	INTELSAT	1984
Venezuala	144	INTELSAT	1982
Zaire	36	INTELSAT	1978

[a]Used INTELSAT facilities prior to using a national satellite system.

Let us first review the present situation of countries using satellites for national communications networks and whose satellites they are using at the present time. These countries are shown in Table 8.3 and also the satellite bandwidth capacity in which they have invested.

Many lesser developed countries are now using satellite communications as a means for developing their economies, educating the people, and providing telecommunications into areas hitherto impossible. This chapter discusses some of the approaches taken so far, including the commissioning of national satellites, and the more modest approach of leasing or purchasing transponders from other organizations such as INTELSAT.

8.1. INTELSAT SUPPLIED NATIONAL SYSTEMS

INTELSAT's charter is contained in two documents known as the Agreement and the Operating Agreement. These agreements, in part, define INTELSAT's ability to provide services for both international and national public telecommunications. Article III(b) of the Agreement states that the INTELSAT space segment may also be used to provide domestic public telecommunications, when domestic services are determined to be:

Between areas separated by areas not under the jurisdiction of the state concerned, or between areas separated by high seas, or

Between areas which are not linked by any terrestrial wideband facilities and which are separated by natural barriers of such an exceptional nature that they impede the viable establishment of terrestrial wideband facilities between such areas, provided that appropriate approvals have been given in advance.

Such domestic services are then considered on the same basis as international public telecommunications services, which is INTELSAT's primary purpose. These provisions within the Agreement mean that the use of those domestic services normally increase the country's investment share, and its voting participation at the organization's Board of Governors. INTELSAT, being a cooperative ownership organization, compensates member countries for their investment, which consequently decreases considerably the net cost of leasing transponders for this purpose, and has made the use of INTELSAT for this purpose attractive to many countries. This cooperative ownership effectively reduces the leasing cost by about 40% when the return on investment is taken into account as occurs under INTELSAT's financial arrangements.

The first nation to make use of these provisions was the United States which leased one transponder to expand services between the mainland States and the State of Hawaii in 1974. These services were later transferred to a wholly United States satellite when one became available. Since that time, of course, many more domestic satellites have come into being for the United States. Algeria followed in 1975 and has operated a very reliable and efficient domestic network on a leased transponder ever since. Brazil and Malaysia followed quickly behind Algeria in the same year, and even at this early stage two distinct requirements for domestic networks emerged.

The United States required fully guaranteed service for which a rate equivalent to a full 36 MHz global beam transponder on a satellite operating in the international network carrying 360 channels was charged. The Algerian, Brazilian, and Malaysian requirements were for leased transponders on spare satellites which could not be fully guaranteed in the event of satellite failure but having a significantly lower charge than that for the fully guaranteed service. This service came to be known as preemptible, and was available to countries on a 5-year lease basis at a greatly reduced cost, initially set at US$1,000,000 per year for a global beam transponder. Pro-rata tariffs were set for fractional transponders with units as low as 9 MHz and multiples thereof. There were similar pro rata cost differences for different types of transponder. This basic US$1 million annual charge was reduced to US$800,000 in 1980, and again to US$680,000 in 1985. The low cost and high reliability of this service proved very popular, and the effects of the 1980 tariff reduction can easily be seen in the growth table. Similar growth increases are expected from the reduction in cost and wider variety of services introduced in 1985–1986.

One of the main reasons why the preemptible leased transponder service has been so popular is that preemption has never been necessary, as INTELSAT for many years had plenty of spare capacity. This situation is changing, but for many years, many countries have clamoured for a domestic service that could be planned with the knowledge that the capacity would always be available. To provide "planned" domestic services, however, necessitates providing additional satellite capabilities with additional costs to be borne by all signatories.

Conversely, the revenue from domestic services was attained from spare capacity that would have been idle without the service. This certainly helped in keeping down the investment needed by the large signatories. The issue has been a point of contention for many years and is now reaching greater interest with the advent of entrepreneurial private companies trying to compete with the international organizations.

INTELSAT's charges for selling and leasing transponders for domestic purposes are shown in Table 8.4. Leasing charges vary with the term and priority of the lease. The increased selling cost of INTELSAT VI transponders reflects the longer life expectancy of the latest satellite series.

INTELSAT studied and discussed the potential domestic service market for almost ten years, not resolving the real issue of whether to provide domestic service in such a way that countries could reliably plan on the service being available on a long-term basis. On different occasions both the Assembly of Parties and the Meeting of Signatories asked that the service be

TABLE 8.4   Synopsis of INTELSAT Domestic Tariffs

Beam Type	Bandwidth (MHz)	Selling (US$M)	Leasing Prices (US$M per Year)	
			Lowest	Highest
INTELSAT V/V-A Satellites				
C-Band				
Hemi/zone	9	N/A	0.170	N/A
	18	N/A	0.340	0.496
	36	2.602	0.680	0.993
	72	3.472	1.360	1.985
Spot	18	N/A	0.714	0.856
	36	4.489	1.428	1.713
Global	9	N/A	0.170	N/A
Ku-Band				
	9	N/A	0.204	N/A
	18	N/A	0.408	N/A
	36	N/A	0.816	1.192
	72/241	5.7	1.832	2.382
INTELSAT VI Satellites				
C-Band				
Hemi/zone	72	9.4	*	1.777
Hemi/zone	36	6.435	*	1.215
Ku-Band				
Spot	150	26.498	*	5.049
Spot	72	14.053	*	2.676

*Not yet determined.

improved while many countries and regions searched for better solutions that could be made available through INTELSAT. Regrettably, proportionately less and less of that demand was being satisfied through INTELSAT because it simply did not provide the kind of service that countries would opt for if they had the choice. This permitted satellite manufacturers and independent consultants to expand the advantages of separate domestic systems and downplay the enormous expenses involved. By 1988, separate national systems developed in several countries. Despite this, the demand for INTELSAT domestic services has not diminished.

Although INTELSAT has introduced some variety in its menu of domestic services with regard to purchase of spare transponder and variable rates and priorities, with reason the basic commodity available is primarily a lower-powered, omnidirectional transponder that requires relatively expensive earth stations. As long as this situation remains, INTELSAT can expect the number of other satellite systems to grow. On the other hand, the choice made by INTELSAT member countries has been deliberate in order to minimize the number of satellites, and keep down the investment cost, as well as the cost of international services. The new INTELSAT VI and VII series will greatly improve INTELSAT's potential for supplying domestic services due to their increased varity of transponders, provided they are attractively priced.

A summary of the present types of service and facilities are shown below.

Transponders in C- or Ku-bands may be leased or purchased outright.

National television programs may also be received by countries at a slightly higher tariff.

Fractional transponders, in units of 9 MHz, in C- or Ku-band may be leased on a preemptible or nonpreemptible basis. Tariffs vary accordingly and also depend on whether the lease is cancellable or noncancellable.

Leases are available for full-time, part-time, or occasional use service.

Each nation may devise its own transmission plan based on its own needs. There are, obviously, technical limitations that protect other users.

INTELSAT provides free consulting services in planning domestic services under its Assistance and Development Program (IADP).

To support nations using its satellites and other facilities, INTELSAT also provides an assistance and development program which covers a range of activities designed to help each nation plan, develop, install, and use the satellite network. The program encourages the active participation of nationals in all its advisory activities which may include:

Initial system and network planning

Viability and feasibility studies

Locating sources of development funds

Preparing the request for proposal (RFP)

Reviewing proposals

Monitoring progress and installation

Introducing training programs

Providing training assistance and documentation

Introducing maintenance, operation, and maintenance management programs

The Program, since 1978, has helped many countries in the implementation of domestic leased or purchased networks.

Having seen the number of countries that have used, and continue to use INTELSAT for domestic purposes, and noting that some countries chose to purchase their own satellite, this seems an appropriate moment to consider why this choice has been made before proceeding to examine some specific examples of countries using satellites for their national telecommunications networks.

8.2. INTELSAT VERSUS NATIONAL OWNERSHIP

The stones thrown in favor of a country launching and operating its own satellite system are that in the long-term it is more economical than using INTELSAT's lower power satellites, arguing that higher power satellites can be used which permit the use of smaller and cheaper earth stations, national security of ownership, and probably the biggest selling carrot of them all—national aspirations.

The ecomonics of owning and operating a national satellite system vis-à-vis using cooperatively-owned INTELSAT satellites are complex. The arguments used to claim that a national satellite is more economical, ususlly assume that the satellite, or satellites, are fully used from launch to the end of life, and also upon earth stations being much cheaper due to the satellite being more powerful. The first point is, of course, a complete fallacy because it indicates that satellite saturation would be immediate and that no growth would be needed for many years. All telecommunication is a growth process, and all national networks grow as the demand grows. The second point that the networks are much cheaper due to higher satellite power is much more complex and there are good reasons why users prefer smaller earth stations. INTELSAT failed to provide the type of satellite needed for this purpose, which is probably why several countries decided to purchase their own. It is interesting to note in retrospect that the INTELSAT IV series was able to provide considerably more down-link power than the later series, and became the base for satellite manufacturers to promote national satellite systems.

There is no question that, on strictly economic grounds, INTELSAT can provide a very much cheaper alternative to almost any country than that provided by a national satellite. There are many reasons for this.

Firstly, a domestic satellite system must be designed to accommodate the most stringent requirements as a total system which could include both public telecommunications and military requirements for national security. Such satellites require individual design and are consequently more costly than the satellites designed for more general use. On the other hand, a large satellite service supplier such as INTELSAT can provide different degrees of reliability and protection as needed on different transponders or, if necessary, rent satellites with very significant economic advantages.

Secondly, any domestic satellite system must be designed to meet all reasonable requirements over a seven to ten year life of that system, but pay for that system before or at launch, which significantly increases real cost. When all the costs are added, a relatively simple domestic system, excluding the earth stations, is estimated to cost between US$300 and US$400 million.

In the case of leasing or purchasing transponders which are available as needed, the cost is very much less. At the present rates, 12 transponders could be leased for ten years at an approximate net cost of US$82 million on a preemptible basis or US$140 million on a nonpreemptible basis.

Thirdly, any domestic satellite system requires a ground-based earth station, or stations, for tracking, telemetry, control, and monitor purposes (TTC&M). These facilities must function 24 hours a day for the operational life of the satellite and impose a large extra cost for the system, whereas the service is included in the cost when obtaining the service from INTELSAT.

With this economic background it is difficult to understand why some less wealthy countries have pursued, and are continuing to pursue, the path of individual ownership. One must assume that the lures of less expensive earth stations, national security, and national pride have prevailed in several cases; cases of individual countries are described in Chapter 9.

One major technical difference between INTELSAT and some of the domestic satellites is the polarization used. INTELSAT II and III series satellites were linearly polarized. The transition to circular polarization was made for the singly polarized INTELSAT III series to avoid Faraday rotation effects and antenna beam polarization alignment requirements. For singly polarized linear systems these effects impacted mainly power loss and satellite point-over capability.

With the introduction of dual polarization and increased frequency reuse on the INTELSAT V series, the use of circular polarization turned out to be quite fortuitous from the viewpoint of the satellite antenna design. Frequency reuse using multiple spatially isolated beams required offset reflector systems on the satellite. Achieving dual circular polarization from offset reflectors of practical focal lengths is straightforward, but it is exceedingly difficult to achieve dual linear polarization specifications from similar offset reflectors with INTELSAT beam coverages. Other satellite systems with considerably simpler beam coverages than INTELSAT achieve dual linear polarization using different gridded or grated reflector antennas for each sense of polarization which, in practice, requires that the transmit and receive beams for a given polarization come from the same antenna system. Formation of transmit and receive band multiple beams from a single large feed array is difficult due to the large relative changes in feed element spacing with respect to the operating wavelength. From the satellite antenna design viewpoint, there are therefore strong reasons to support the continued use of circular polarization at C-band if extensive multiple beam frequency reuse continues to be used by INTELSAT.

For the earth station antennas, dual circular polarization requires good quarter-wavelength polarizers which become a serious cost factor particularly for small antennas. However, dual linearly polarized systems with a requirement for arbitrary polarization alignment capability for accommodating different satellite locations and rapid point-over capability, are quite likely to use rotatable half-wave polarizers with a similar cost impact for small antennas.

INTELSAT needs to service a large number of earth stations at low elevation angles which requires satellite antennas providing wide or global beams. The question of circular versus linear polarization needs to be considered from an overall system point of view and the disadvantages of dual linear polarization for INTELSAT at C-band are briefly that such design:

Complicates the satellite antenna design if multiple beam frequency reuse beam coverages are required. This is necessary to obtain maximum use of the available orbital resources.

Complicates earth station point-over to spare or other satellite locations; the polarization alignment of the earth station must be adjusted whereas readjustment is not necessary for dual circular systems.

Use of a half-wave rotatable polarizer system for dual linear polarization has at least as much cost impact on small terminals as the quarter-wavelength polarizer requirements for dual circular systems.

Faraday rotation effects become a factor for dual linear systems which can be just ignored, periodically adjusted, or polarization-track the Faraday rotations with different attendant levels of impact on the polarization isolation achievable and the cost and complexity of implementing and operating the system.

INTELSAT, at an early stage, chose the path of circular polarization of C-band when the III series was introduced, and for the sake of uniformity, this has continued.

8.3. TYPICAL NATIONAL TELECOMMUNICATIONS NEEDS

Each country has its own unique communications problems. The domestic services provided by INTELSAT to 34 countries can be grouped into four major catagories:

Type of Service

Trunk telephony between cities
Thin-route telephone in remote areas
Television distribution
A combination network

Although telephone may seem the most basic requirement, television is obviously considered equally necessary for national news and education. Some typical transponder configurations are shown in Figures 8.2 and 8.3. These figures illustrate the fact that almost every national network needs both telephone and television.

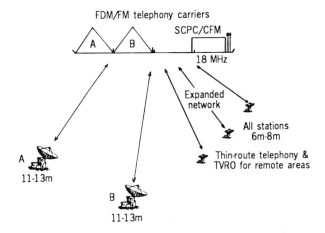

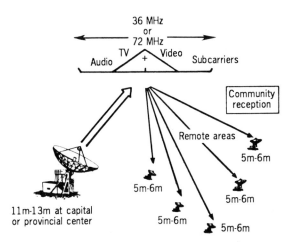

Figure 8.2. Typical large domestic networks.

Trunk telephony services are required between larger towns and require relatively large earth stations in the order of 10–11 m diameter antennas. The satellite network may create or complement existing terrestrial facilities by carrying new or excess traffic, or providing diversity routing. In many cases, the new services will be the first available service and can be expected to develop quickly. FDM/FM carriers were used earlier but more attractive transmission techniques are now being used more and more, such as companded FM, SCPC, or digital means.

INTELSAT V, 9 MHz, global beam lease

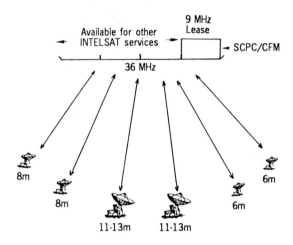

INTELSAT V, 36 MHz, global beam lease

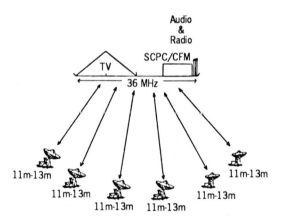

Figure 8.3. Typical small domestic networks.

Thin-route telephony services are designed to connect a large number of remote locations with the larger towns. While each location has only a few telephone circuits, these may aggregate into a fairly large national requirement depending on the geography of the individual country. The remote locations use relatively small earth stations in the order of 5–6 m diameter antennas and almost universally use SCPC techniques to carry their voice

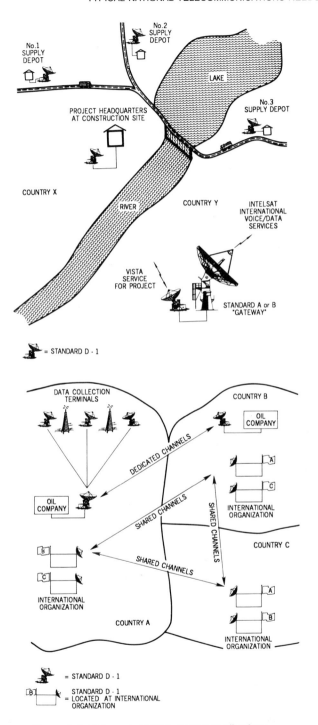

Figure 8.4. Thin-route VISTA service applications.

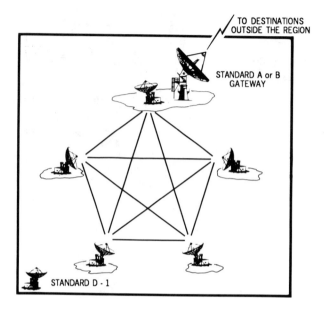

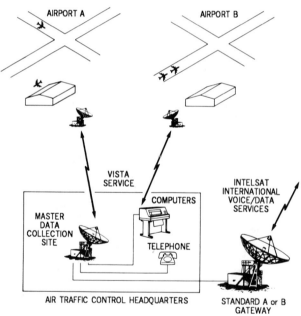

Figure 8.4. (*Continued*)

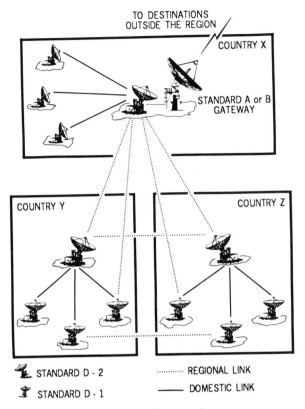

Figure 8.4. (*Continued*)

and low speed telex or telegraph traffic. Depending on national policy, television receive capabilities can also be installed at these earth stations.

Television distribution services can be transmitted over a large area by satellite from a central broadcasting facility to smaller earth stations either for rebroadcasting via a VHF/UHF transmitter or for direct reception by local communities. The received signal needs to be of a higher quality if it is to be retransmitted than if it is to feed a television set directly. Rebroadcasting necessitates larger antennas for reception purposes. For INTELSAT satellites, the antennas are usually about 5 m in diameter for community reception, and about 10 m for rebroadcasting.

Some other typical requirements for national networks include a network for natural disasters or emergency situations, capacity to back up terrestrial microwave or cable systems, special development projects, government security networks, and possibly aeronautical, maritime and mobile applications.

INTELSAT's domestic services allow each country to have its own earth station design and transmission planning. The combination of earth station network and satellite resources can be optimized to meet individual requirements provided that all interference criteria are met. The interference criteria applied to both C-band and Ku-band domestic earth stations are provided by INTELSAT in its Standard Z earth station performance characteristics. Examples of the mandatory performance characteristics are off-beam emission power density, transmit voltage axial ratio, and transmit power stability. Other earth station characteristics such as G/T, channel quality, and modulation technique may be chosen by the individual country.

It is also possible for a country with a very small requirement to use INTELSAT's VISTA service which was introduced in 1983 to permit smaller and less costly earth stations where traffic requirements are very small. This service permits space capacity to be leased on a per-channel basis using the lowest possible cost earth station and equipped with just a single or just a few channels. There are many ideal applications for this service only a few of which are shown in Figure 8.4.

8.4. SMALL NATIONAL NETWORKS

Many countries have a small requirement for domestic satellite communications that, without any television requirements, does not warrant the lease of even 9 MHz space segment capacity. These needs are for low density telephone service for rural or remote areas which require very low cost earth stations. Two cases that were studied by INTELSAT during the early 1980s are typical of this type of requirement.

Firstly was Ecuador with a need for a link of less than 12 telephone channels between the country's mainland and the Galapagos Islands. Existing communications were by high frequency radio which required improvement as well as increased capacity. Secondly, was the South Pacific region which consists of a number of small, newly emerged separate nations having only minimal telecommunications facilities and, which had for a number of years, relied heavily upon an old experimental satellite provided by NASA.

The first case was clearly a domestic issue and services could be provided by INTELSAT in a number of ways under existing services. Standard telephone service would have been possible using a Standard B (10 m) earth station in the Galapagos working to the country's Standard A earth station of the mainland. INTELSAT was reluctant to permit this type of service on the Atlantic Primary satellite which was heavily congested. The alternative was for Ecuador to build a new Standard B earth station both on the mainland and another on the Galapagos Islands. They could either lease 9 MHz of satellite capacity which would accommodate approximately 50 telephone circuits or lease individual telephone circuits individually. The two earth stations were estimated to cost about US$1.5 million, and the project was considered to be too costly.

The second case of the South Pacific had both domestic and international requirements and very few of the countries concerned were members of

INTELSAT. The economies of the small island nations were, and still are, such that any expenditure of this nature would be too costly. Several services had been provided by NASA's ATS-1 satellite to small island communities and such institutions as the University of the South Pacific, but it was clear that most of these were unwilling and unable to pay for the services at standard rates through their telecommunications administrations. Many studies were performed by various humanitarian organizations at great cumulative expense and all concluded that satellite communications would be the only solution. The real question then became which satellite system would be used—INTELSAT, a South Pacific Satellite, or a satellite system using facilities leased from countries such as the United States, Japan, Australia, or Indonesia.

With these two cases and many similar ones in mind, INTELSAT reviewed the world wide requirements for this type of service in 1981 and at the end of 1983 introduced what became known as the VISTA Service.

VISTA was created to provide basic communications facilities for rural and remote communities having inadequate or no telecommunications facilities. Typically, VISTA would be used by a number of countries each with a relatively small requirement arising from isolated or remote communities. Domestic and international telecommunications are permitted and efficient use of the space segment is achieved by consideration of individual requirements and effective sharing of the satellite capacity. Space segment circuit costs were introduced at the low cost of US$6360 per year.

Small, low cost earth stations are used with simple access and signaling to operate in which have become known as Star or Mesh, or combined configurations are shown in Figure 8.5. Two new earth station standards were introuced for the VISTA service—the Standard D1 and Standard D2. The small, 4.5 m, Standard D1 has less stringent characteristics than any other INTELSAT earth station and is intended for the small remote locations. The Standard D2 is very similar to the Standard B station having an antenna of about 10 m diameter and is suitable for the larger, hub earth stations. Either SCPC/CFM or SCPC/FM transmission techniques are used or without demand assignment, but each channel should normally be voice activated. The voice activation requirement may be overlooked in cases where low speed data, telex, or facsimile are used.

Although introduced in 1983, the service has been very slow to develop due to the cost of the earth stations. Although basic earth station costs were first estimated to be only about US$25,000, these earth stations were far too rudimentary for practical purposes and the cost rose to nearer US$100,000 each when essential redundancy, spare equipment and installation were taken into account. INTELSAT later tried to overcome this problem by itself offering to act as purchasing agent for bulk purchase of these earth stations which would allow reduced costs to countries buying only a few of them. At the end of 1988 there were still under 60 VISTA channels being used in the whole INTELSAT system. However, I believe there is an

"STAR" NETWORK

"MESH" NETWORK

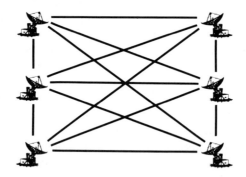

Figure 8.5. Star and Mesh network principles.

enormous worldwide potential for this type of service, and with lower cost earth stations, the growth of VISTA will escalate quickly.

Future expansion must be considered when planning any VISTA system, as in any other telecommunications service. In the case of VISTA, it may prove very difficult, expensive, or even impossible to expand earth station facilities without complete replacement of the earth station. Some of the most important things to consider are:

Can additional channels be added in the future economically?

At what expansion level would new power amplifiers be required?

At what expansion level would a new, larger earth station be required?

Could the earth station be easily moved to another site?

Could the earth station be modified at some later date to accommodate a dual polarized feed system with a better axial ratio which would be necessary if they transferred to a full domestic transponder service?

Is television reception likely to be a future requirement?

Although VISTA was introduced as a basic telephone service, the future need for receiving television could be a rapid follow-on requirement to the basic VISTA service. Television is becoming a common service all over the world, and many countries are already using it for education, health, and information purposes. If the original basic earth stations are likely to be needed to receive television, it is certainly much less costly to make this modification easy from the start.

As mentioned earlier, VISTA networks have been slow to develop due to earth station costs, but likely applications that are developing include:

Oil exploration in remote places
Thin-route networks, national or international
Small domestic networks
Small regional networks
Special development projects
Diplomatic networks

9

NATIONAL SATELLITE NETWORKS

In 1989 there are approximately 50 countries using satellites for providing some, or all, of their national telecommunications or broadcasting services. Some own and operate their own satellites; some share regional satellite systems; others make use of international cooperatives or other country's satellites. This chapter provides a brief description of each of these countries' satellite services and how each reached their present position. Many countries' uses of INTELSAT are very similar, and the information on some countries is difficult to obtain, so this chapter contains general and typical examples of how some 50 counries make use of satellites for national telecommunications purposes.

ALGERIA

As mentioned earlier, Algeria was the first country to make use of the INTELSAT Agreement for domestic telecommunications and made its intentions within a few months after the definitive agreements came into force at only the third meeting of INTELSAT's Board of Governors in July 1973. Satellite capacity at that time was very small compared to the present time, and it was decided that services of this nature could not be provided on the primary satellite but that they would best be provided using spare satellite capacity. This immediately raised questions of charging policies and how such services should be charged. The outcome was that Algeria leased a 36 MHz global beam transponder on a preemptible basis at a lower annual rate set at US$1 million. The first ten years of satellite capacity cost Algeria less than US$9 million as rates were later reduced. This was reduced to about US$6 million taking into account INTELSAT's financial arrangements.

To support its use of INTELSAT capacity for domestic purposes under Article III b(ii), Algeria had to describe the geographic nature of the country and why satellite communications were the only viable means for providing telecommunications within its borders. The Algerian description of its proposed domestic system became a model for all future applications of this nature. Geographically, it is also very similar to many other countries, some of which still do not use satellites.

One of the larger countries of the world, Algeria covers an area of 2.4 million km^2 and has a population of over 10 million. A map of Algeria is provided in Figure 9.1. The Saharan regions served by domestic satellite telecommunications services occupy over 75% of the total territory of Algeria. That region is in the shape of an irregular hexagon which lies south

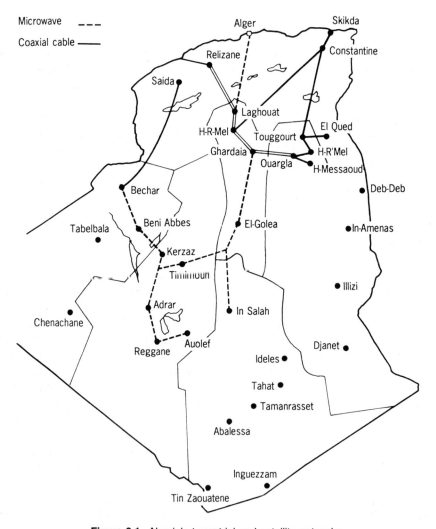

Figure 9.1. Algeria's terrestrial and satellite networks.

of the city of Laghouat, and has sides ranging from 800 to 1500 km in length. The northern part of the country, a Mediterranean or semidry area of about 360,000 km^2, includes over 80% of the country's population, while under 3 million people live in the Saharan regions, generally clustered around isolated oases which are separated by long stretches of desert covered by large drifting sand dunes, rocks, or mountains such as Tassili, Hoggar, and Tadmait.

The various Saharan settlements are quite far from each other and are difficult to reach by road which, more often than not, is nothing more than a treacherous and barely drivable trail. Thus, Bechar is located 700 km from the Mediterranean port of Oran, the nearest coastal city, while Adrar is 600 km to the south and Tindouf 900 km west of Bechar. In the eastern part of the Sahara, the Algiers–Ouargla road stretches over 800 km, while Tamanrasset is located 1500 km south of Ouargla. The city of Djanet, which, for all practical purposes, can only be reached by air, is locked in a rugged rocky and mountainous area 500 km northeast of Tamanrasset; it is preferable to travel the 1500 km via In-Amenas which is linked with Ouargla by a hilly paved road.

Size and distance considerations notwithstanding, the Sahara is especially known for its inhospitable climate. It is one of the hottest and most arid areas in the world. Summer temperatures often reach 50°C in the shade. In the sun, buildings, and equipment are exposed to temperatures which often reach 65–75°C. The daily temperature variation is considerable, above 50°C, causing rocks to crack at night due to the cold. On the average, the Sahara receives less than 100 mm of rainfall per year, and there are areas with less than 10 mm (Reggane, 8 mm). Rainfall is also irregular and there may be no rain at all for serveral years, and when it does eventually come it may take the form of a real deluge destroying trails and houses. Drought periods can last from three years in the Plateau of Tadmait to five years in the Hoggar.

Violent windstorms are very frequent in the Sahara with velocities of up to 200 km/hour. Devastating sandstorms are a routing occurrence. They stop traffic, paralyze all activities, and erode materials and equipment, without mentioning their harmful effects on radiowave propagation conditions. In addition to power outages, it is not unusual to have link outages or degradations lasting several days or even weeks.

These geographic and climatic conditions, combined with the environment, constitute a natural barrier of an exceptional nature for terrestrial wideband facilities, not only because they require performance characteristics which are especially difficult to meet in terms of equipment, buildings, and power sources, but also because they raise practically insoluble operational and maintenance problems.

Based upon this description of its geographic nature, it was determined that not only did all Saharan settlements lack adequate and high quality telecommunications facilities, but that in addition, the implementation of wideband terrestrial facilities was either impossible to accomplish or would be a lengthy and costly process, requiring considerable and expensive operational and maintenance activities.

During the first ten years of satellite communications, Algeria did extend its terrestrial microwave system to provide backup facilities where it was practicable. It was therefore possible to look back and consider some of the advantages of satellites over terrestrial means in Algeria.

The choice of a satellite network over a microwave network providing equivalent services, assuming that such a network were to be fully achievable, was dictated at the time by such considerations as:

Satellite costs were one eighth of equivalent terrestrial costs

Lead times were three to four times shorter

Maintenance was greatly reduced because earth stations, located in towns, needed fewer technicians and vehicles than microwave relay stations scattered over 7000 km

The network was much more flexible

Transmission stability and quality were greatly reduced

Site security was much greater

Algeria's domestic satellite system has been extremely successful in developing its growth and resources. The original 15 earth stations which were supplied by the United States' GTE Corporation, have been increased to 37, and a list of earth stations is provided in Table 9.1. Plans are being made to increase the satellite capacity and to expand the network further.

Algeria estimated that a similar expansion using terrestrial microwave would have involved more than 350 relay stations at a cost of US$120 million as well as taking years to install. By comparison, an earth station

TABLE 9.1   Algeria's Earth Stations

First phase	Year	Second Phase	Year
Lakhdaria (Master)	1975	Abalessa	1986
Adrar	1976	Aoulef	1986
Ain-Amenas	1978	Bordj Omar Driss	1986
Bechar	1975	Chenaached	1986
Beni-Abbes	1977	Debdeb	1986
Djanet	1976	Gara Djebilet	1986
El-Golea	1977	Ideles	1986
El-Oued	1977	Illizi	1986
Ghardaia	1976	Inguezza	1986
In-Salah	1976	Kerzaz	1986
Ouargla	1975	Oumelassel	1986
Reggane	1979	Tabelbala	1986
Tamanrasset	1976	Tazrouk	1986
Timimoun	1976	Tin Zaouaten	1986
Tindouf	1976	Touggourt	1986

network expansion could be accomplished in two years and cost only US$20 million.

ARGENTINA

Argentina hosted the World Football Championship (World Cup Soccer) in 1978 which motivated the country's movement towards satellite communications resulting from its lack of facilities to broadcast the games by television either nationally or worldwide. Shortly after the end of the games, Argentina requested INTELSAT to start leasing one and a half global beam transponders starting in 1982, and in order to develop the necessary earth station network, a loan of US$27.75 million was obtained from the U.S. Export-Import Bank to purchase 38 earth stations from the Harris Corporation of Melbourne, Florida.

The country has a surface area of nearly 3 million km^2 which is almost the size of the whole of Europe excluding the Soviet Union. Its population, however, is under 30 million and is not evenly distributed. About one third of the population lives in the region of the country's capital, Buenos Aires. The economy is predominantly agricultural in the form of grain and cattle with large and widespread farms and ranches. Argentina has its own petroleum and energy resources and thus has no major energy problems.

In 1982 the country had a telephone network of about 2.5 million telephones, 6000 telex connections, and more than 10,000 km of wideband terrestrial connections, and more than 10,000 km of wideband terrestrial trunks which had grown radially from Buenos Aires. These links had grown up along the nation's railway system towards the major towns and national borders and often missed out the smaller towns and villages. In 1876, only four years after Alexander Graham Bell invented the telephone, there were 20 telephones linked to a central exchange in Buenos Aires.

Telecommunications in Argentina are owned and operated by the Empresa Nacional de Telecommunicaciones (ENTEL), which is an independent company responsible to the country's Ministry of Communications which has been an INTELSAT Signatory since 1965. An international earth station at Balcarce, near Buenos Aires, has been operational since 1968.

The national satellite system, using transponders leased from INTELSAT, was planned to support the existing terrestrial network and also to integrate the population not covered by the existing network. In addition to expanding the phone system, the whole country was to be supplied with color television and radio broadcasting for the first time. ENTEL planned a system that would include emergency backup for failures in the terrestrial trunk links, and that would also allow future growth and flexibility necessitating the minimum maintenance and supervision. The terrestrial and satellite networks at the time that satellite services were introduced in 1982 are shown in Figure 9.2.

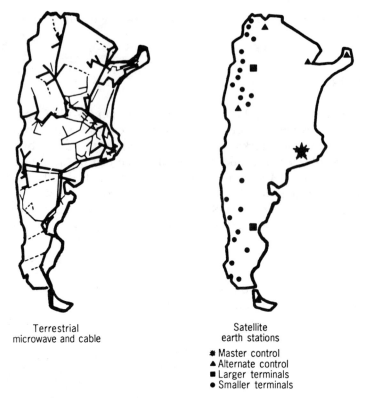

Terrestrial
microwave and cable

Satellite
earth stations

* Master control
▲ Alternate control
■ Larger terminals
● Smaller terminals

Figure 9.2. Argentina's terrestrial and satellite networks.

Argentina is fully covered by the hemispheric beams of INTELSAT satellites, and although higher power steerable spot beams were being planned by INTELSAT at that time, design of the system and of the earth stations was based upon the INTELSAT V hemispheric C-band transponders. Television was to be carried using a 36 MHz transponder, and the telephone service on an 18 MHz segment of another transponder.

The telephone service channels each use frequency modulated SCPC radio carriers separated by 45 kHz on a demand-assigned basis permitting about 180 simultaneous telephone circuits in the 18 MHz leased transponder segment. Digital scrambling was also included for government use. A list of earth stations being used in the national network is given in Table 9.2.

Argentina is considering the procurement of its own national satellite system and filed notice in 1985 that it requires two geosynchronous orbit locations at 275° and 280°E. The satellites would be called "Nahuel" and the cost would be about US$240 million (1983 estimate) for launching in 1987 or 1988. At the time of writing no firm plans have been made for the purchase or deployment of these satellites, following the launch and insurance problems that came about during 1986.

TABLE 9.2   Argentina's Earth Stations

Antarctica 1	Las Lenas
Antarctica 2	Los Antiguos
Antarctica 3	Los Menucos
Antofagasta	Mendoza
Bacarce 3 (Master)	Neuquen
Barreal	Paso de Indios
Butaranquil	Perito Mereno
Camanas	Puerto Iguazo
Chepes	Resistencia
Comodoro Rivadavia	Rinconada
Coranzuli	Rio Grande
El Calafate	Rio Mayo
Fiambala	Rio Turbio
Gan Gan	San Antonio de Cobres
Gobernador Gregores	San Salvador de Jujuy
Ing. Jacobacci	Susques
La Rioja	Vinchina
Lago Puhelo	5 Transportables

Argentina purchased one 72 MHz west hemispheric transponder and two 36 MHz global beam transponders on INTELSAT V, F-3 at 307°E in 1987. These transponders were later replaced by three C-band spot beam transponders and a west hemispheric beam transponder on the replacement satellite, INTELSAT V-A, F13 in 1988 which allowed Argentina to expand its television programs much more effectively.

AUSTRALIA

Australia is a country of nearly 8 million km^2 and about 15 million people divided into six states, plus the Northern Territory, the Federal Capital as well as the large island of Tasmania, numerous islands in the surrounding ocean regions and part of Antarctica. Nearly half the country is tropical, and climates vary considerably, ranging from arid desert in the interior to very high rainfall on the northern coast.

Population is concentrated on the southeast coast with only about 5% living in the tropical northern half of the country. The small rural population is scattered in small, isolated, communities involved in primary industries such as agriculture and mineral extraction. Telecommunications could not be provided to these rural locations in the same way as urban areas prior to the arrival of satellite communications and maps of Australia showing the population density in 1976 and the wideband terrestrial network in 1980 are shown in Figures 9.3 and 9.4.

Australia is a perfect example of a large sparsely populated country in which communications satellites can, and do, play such an important role. It

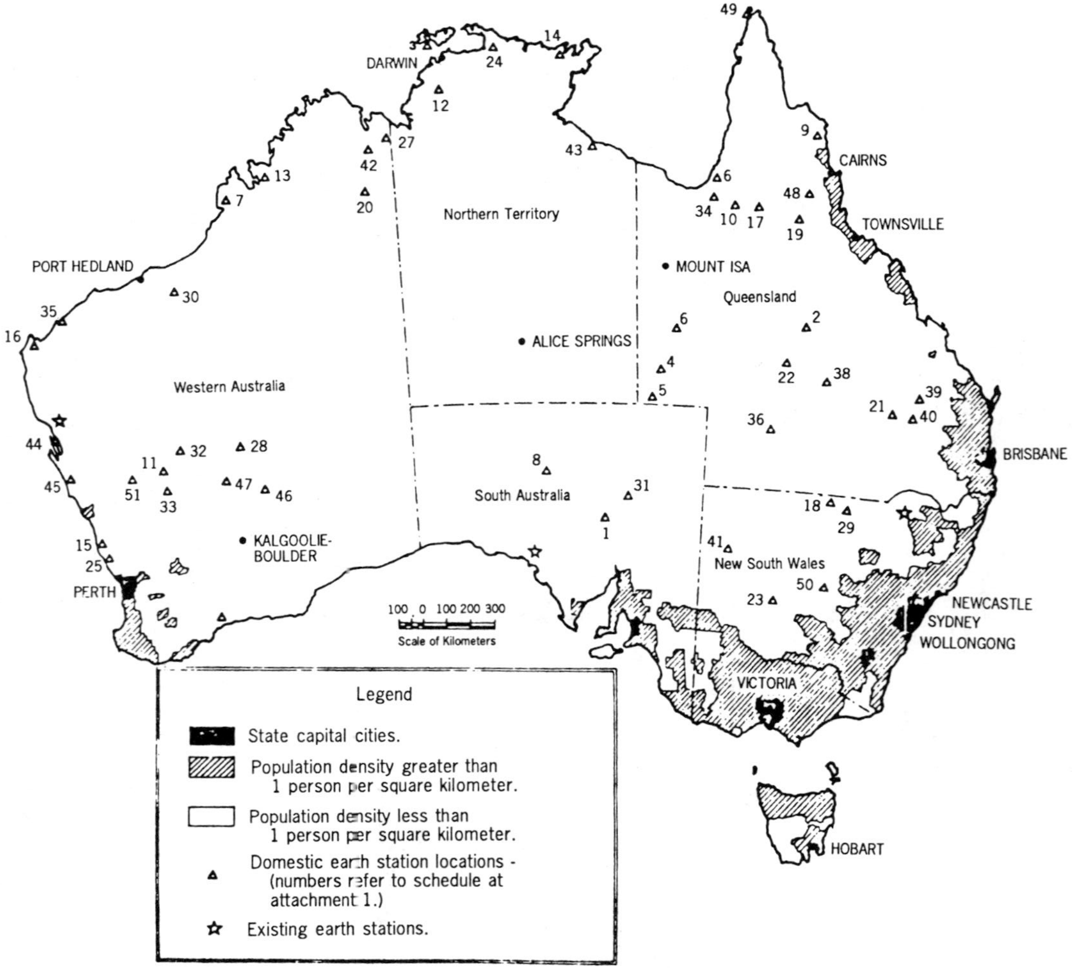

Figure 9.3. Australia's population spread in 1976 (courtesy: Australian Government Publication)

Figure 9.4. Australia's broadband network in 1980 (courtesy: Australian Government Publication).

Legend

—— Broadband routes (microwave and coaxial cable) by 1980. Thicker lines indicate two or more routes provided.

—— Domestic TV only routes

▲ Domestic earth station locations - (numbers refer to schedule at attachment 1.)

☆ Existing earth stations.

100 0 100 200 300

Scale of Kilometers

193

is also a case study that has been extraordinarily well documented and its new AUSSAT satellite system inaugurated in 1983 promises to be one of the most economically viable domestic satellite networks although at the moment it is still far from profitable. (Edgar Harcourt, *Taming the Tyrant*; Kay Turnbull, *AUSSAT*.)

Telecommunications have always been of major importance to Australia due to its distance from other major trade areas of the world. The first telegraph link in Australia was installed in 1854 between Melbourne and Williamstown in Victoria. In 1878 Australia introduced its first telephone link in South Australia between Semaphore and Port Augusta, and within two years telephones were operating in the major cities. Telegraphic contact with the rest of the world was by telegraph cable which started in 1872.

Australia was one of the founder members of INTELSAT in 1964, and was part of the first major satellite networks used for the U.S. Apollo Project. Since that time Australia has participated heavily in all of INTELSAT's activities. INTELSAT has provided about 50% of all the country's international links with the other 50% being provided by submarine cable systems. Australia has also used INTELSAT for domestic services such as remote area television distribution, but while successful use was made of INTELSAT, the desire to develop an Australian space aerospace industry and a longer term arrangement led to the introduction of Australia's own satellite system in 1985.

Although numerous private studies had been performed, the most important feasibility study to examine the use of satellites was conducted in 1978 by the interested government agencies—the Overseas Telecommunications Commission Australia (OTC), the Australian Department of Communications, the Australian Broadcasting Commission, and TELECOM, Australia. The outcome of the study was that Australia decided to develop its own satellite system and would lease capacity from INTELSAT until their own satellites were launched and operational. To own and manage the system, AUSSAT Pty. Ltd. was incorporated by the government with the intention of selling 49% of the stock to private enterprise.

As an interim measure, OTC leased transponders from INTELSAT which permitted the start of television distribution to all parts of the country between 1980 and 1985.

The satellite planning, started in 1979, was performed by a Satellite Project Office (SPO) whose primary functions were to:

Examine the broad policy implications
Consult with prospective users and determine their needs
Develop a conceptual satellite system
Establish technical and procurement specifications

The SPO drew its expertise from all interested government agencies but OTC provided the bulk of the technical expertise. COMSAT General of the

United States and the European Space Agency (ESA) were employed as consultants during the early planning stages in order to ensure that as much technology transfer took place as possible.

The system's technical specifications were released in October 1980, and OTC was appointed interim owner and manager of the satellite system whose first task was to call for worldwide tenders and their evaluation. Tenders were received from the United States' Hughes Aircraft Company and Ford Aerospace as well as British Aerospace on behalf of European industry.

In November 1981, Australia's satellite operating company, AUSSAT Pty. Ltd. was incorporated and the new company completed the tender evaluation and contract negotiations. It was finally decided to purchase three HS-376 satellites from Hughes Aircraft together with ground control facilities and some earth stations. AUSSAT also at this time completed its review of launch alternatives and contracted with NASA to launch two of the satellites on the space shuttle. The frequencies and beam coverages are shown in Figure 9.5.

AUSSAT was designed to serve the full range of communications needs for the whole country both urban and rural which include:

Low power direct broadcasting for radio and TV

Telephone, including remote homesteads

Voice and data for mining and oil exploration

School of the Air education network

Aeronautical services

Television programming and distribution

Royal Flying Doctor Service

Transportable services for emergencies

Computer and data services for businesses

To provide all these services, AUSSAT also had to develop several types of earth stations to meet the requirements which consist of five major types of earth stations:

Small, low cost receiving stations for TV and radio of about 1.3 m antenna diameter

Small, 2.4–3.0 m, antenna earth stations for SCPC voice/data rural use

Medium size, 4.5–6.0 m, TVRO earth stations for receiving and redistributing TV and radio

Medium size, 4.5–6.0 m, transmit and receive earth stations for multichannel telephone services

Large, 8–18 m, earth stations for larger towns which would form the system's backbone network.

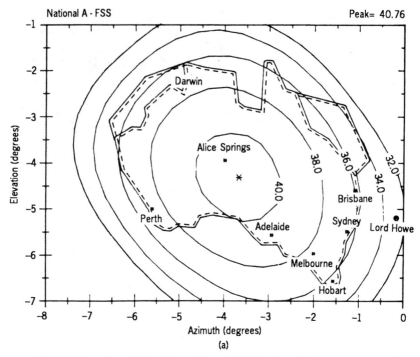

Typical eirp contours (dBW): National A beam (12 Watt channel)

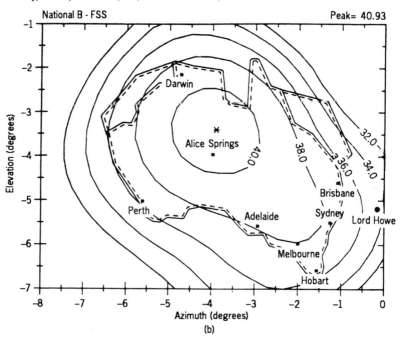

Typical eirp contours (dBW): National B beam (12 Watt channel)

Figure 9.5. Australia's AUSSAT beam coverages (courtesy: Australian Government Publication).

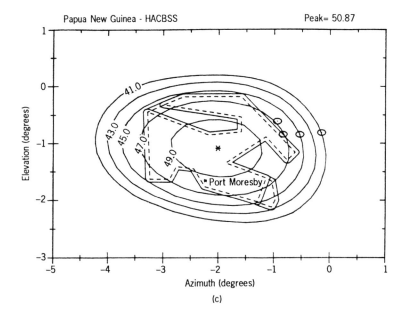

(c)

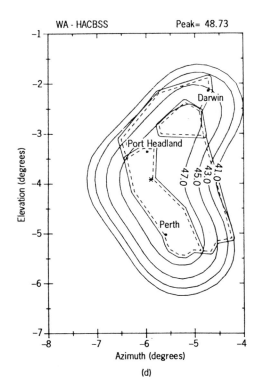

(d)

Figure 9.5. (*Continued*).

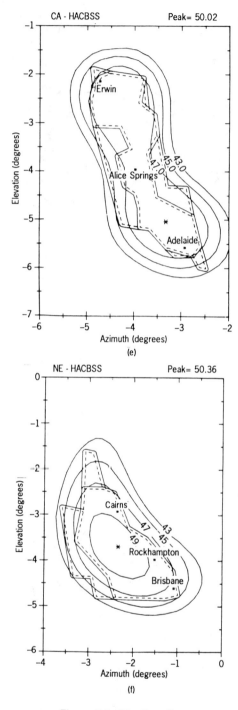

Figure 9.5. (*Continued*).

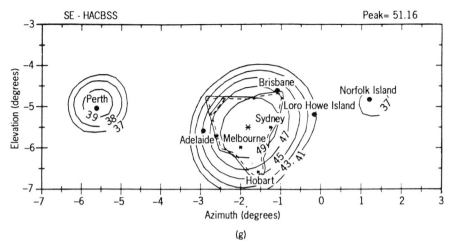

Figure 9.5. (*Continued*).

Many of these earth stations were operational using the C-band trans-ponders leased from INTELSAT and were all ready to give the AUSSAT K-band system a good commercial start. AUSSAT, although still less than 4 years old, is already providing Australia with a much wider range of services than had previously been possible; the satellite is providing enormous benefits. However, in 1986–1987 the satellite company reported a US$35 million pretax operating loss showing that the system is still neither commer-cial nor profitable. At the end of 1987 over 80% of capacity was already in use, and a total of 42 out of a possible 45 transponders are predicted to be in commercial service by late 1988. New Zealand is also leasing capacity on AUSSAT.

In order to provide basic telecommunications services to its remoter territories, Australia is using INTELSAT's VISTA service to Cocos Keeling Island, Christmas Island, and bases in Antarctica. AUSSAT 3 will provide a wider coverage for the off-shore islands, including New Zealand. The second series of AUSSAT satellites is already being planned to replace the present series, and a call for bids was made during 1987 for three more replacement satellites in the mid-1990s that will have higher capacity and power. The five contestants for the second series were Ford Aerospace, Hughes Aircraft, MBB/Aerospatiale (a French–German consortium), RCA, and SatCom International (a British Aerospace–MATRA consortium).

AUSSAT selected Hughes Aircraft for its US$500M contract for two satellites delivered into orbit in 1992. Each will have 15 Ku-band trans-ponders, L-band capacity for mobile services, an experimental 28 GHz Ka-band beacon, and a laser reflector for satellite location keeping.

BRAZIL

In 1960 the telecommunications in Brazil were in very poor shape. The municipal, state and federal governments could all make their own policies, legislation, and their own arrangements for telecommunications. Almost no services were provided by the government which relied entirely on concessions with foreign companies. The lack of central policy or planning led to the government investing almost entirely in foreign companies which again led to little local advantage.

In 1962 a Telecommunications Code was approved by law which represented the beginning of radical change in the national telecommunications within Brazil and to the outside. The main points of the code were that a central agency was established to create a national policy, a national telecommunications company called EMBRATEL, and a national telecommunications fund to support these. The fund was derived from local service taxes and also long distance and international service taxes all of which were controllable from within the country.

EMBRATEL was formed in 1965, and in 1966 the Ministry of Communications took over the greater political authority and control. EMBRATEL then became the national telecommunications administration for Brazil and by the early 1970s had set up a national system in which, at least, the main state capitals had been interconnected, and had started a national and international telecommunications structure using a combination of both terrestrial and satellite systems. There still remained the need for communications to the much smaller and remote towns of this very vast country. To this end, a new organization (TELEBRAS) was created to be responsible for the remaining telecommunications services through EMBRATEL and other operating companies. TELEBRAS was made responsible for planning and promoting public telecommunications services, managing the government's shareholding in such companies, coordination and planning, as well as other interests such as technical planning, training, and overall control of these companies.

As well as EMBRATEL and TELEBRAS, there was also RADIOBRAS which was responsible for broadcasting and television with its own objective of extending its services to all parts of the country. By 1972 there were nearly 2000 municipal districts not covered by even basic telephone service, but by 1982 over 4000 smaller cities and municipal districts were connected into the national system. As far back as 1975, Brazil had notified INTELSAT that it intended to create its own satellite system eventually, but in the meantime it would lease capacity from INTELSAT. INTELSAT was only able to provide wide beam transponders with relatively low power and not on a permanent or planned basis.

Brazil first started leasing capacity from INTELSAT in the same year (1975). The lease was initially a 36 MHz global beam transponder with further increases in 1979, 1980, 1981, and 1982 by which time seven 36 MHz transponders were being leased. It was then decided to launch a national satellite, but Brazil continued leasing the seven transponders until 1985 when the satellites were launched. During these years the network expanded to 33 earth stations by 1985 covering almost all areas of the country with telephone, telex, and television services.

The satellite segment cost for these leases was approximately as follows

Years	Transponders (36 MHz)	INTELSAT Charge (US$)
1975–1978	1	4,000,000
1979	3.5	3,500,000
1980	4.5	3,600,000
1981	4.75	3,750,000
1982	7	4,817,000
1983	7	5,600,000
1984	7	5,600,000
1985 (1st quarter)	7	1,400,000
Total		32,267,000

These are gross figures and do not reflect the refunds due to Brazil for the use of Brazil's capital by INTELSAT. When this is taken into account the net cost of leasing for the same period would have been reduced to around US$20 million for the ten year period. The two BRAZILSAT satellites for the second ten year period cost about US$300 million including launch costs, insurance, and retrofitting the earth stations for linear polarization.

The Brazilian satellite system had been under consideration for many years before it came to fruition. The invitation to satellite manufacturers for the Sistema Brasiliero de Telecommunicacoes por Satellite (SBTS) was canceled in 1977 on economic grounds. A study group later reaffirmed the value of a purely Brazilian satellite system and the project was restarted in 1981 with the two satellites being finally ordered from Canada's Spar Aerospace in June 1982 which involved trade offsets guaranteeing large purchases of Brazilian products.

Each satellite has twenty-four 36 MHz transponders each having a down-link power of 34 dBW compared with the 26 dBW provided by the leased transponders from INTELSAT. However, Brazil had leased seven trans-ponders and now owned 48 transponders; the most optimistic predicted the need for 11 transponders for the 1980s. The satellite design and the shape of its antenna coverage was designed specifically for the contours of Brazil making its use for neighboring countries very limited.

The second series is now under consideration for 1992 which would be another two satellite system but having both C-band and Ku-band trans-ponders as well as wider beams for expanded South American coverage. Figure 9.6 shows Brazil's long distance networks in 1986—an incredible leap forward from what was possible in the 1960s.

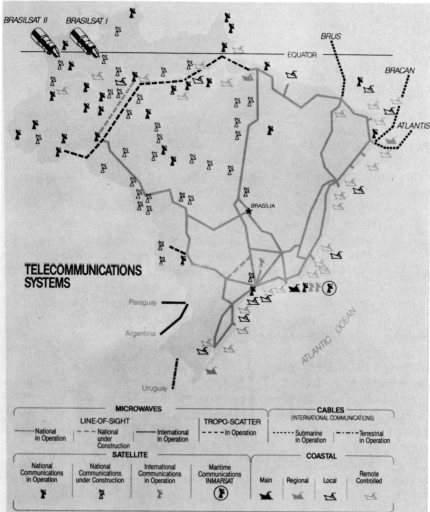

Figure 9.6. Brazil's long distance network in 1965 and 1986 (courtesy: EMBRATEL).

CANADA

Canada's commercial domestic communications satellites are called ANIK (Eskimo for Brother) and provided Canada with the World's second domestic satellite system in 1974, known as TELESAT (The U.S.S.R.'s Molniya system in 1965 was the first).

The ANIK A satellites were similar to the INTELSAT IV satellites differing only in that they possessed more Canadian-built components and the beams were specially designed for the contours of Canada viewed from their orbital locations of 246°, 251°, and 256°E. The three satellites were interchangeable and each provided 12 transponders having either one television channel or 960 voice circuits for their full design life of seven years. They provided Canada with its first national communications system having truly nationwide coverage. Transportable, semipermanent earth stations were developed to meet the demand for services in remote locations and Eskimo villages above the Arctic circle. These earth stations had 4-m antennas and weighed less than 1000 kg complete with shelter. ANIK A also provided a new Frontier TV Service to bring radio and television to small communities that previously had not been part of the national networks.

Obviously these new services would not be immediately self-supporting and were thus not attractive to entrepreneurs. The Canadian government realized the long-term gains to be made by providing good telecommunications services to the whole country, and created and funded the Northern Communities Assistance Program to provide basic local and long-distance telephone service to 28 of the remotest communities of the Northwest Territories. This first ANIK system established Canada as an internationally acknowledged leader in domestic communications via satellite, and the subsequent United States' Western Union WESTAR satellites were modeled on the TELESAT approach, as was the Brazilian system some 10 years later. A map of the early TELESAT system is given in Figure 9.7.

As the TELESAT system developed it was used for new developmental communications covering educational, broadcasting, intercommunity exchanges, and telemedicine whereby doctors could provide advice over vast distances using video and speech to help their diagnosis.

The second series, a single ANIK B, was manufactured by RCA and when launched in 1978, provided a Ku-band transponder and 10 C-band transponders. The Ku-band transponders were leased from TELESAT by the Department of Communications for a series of pilot projects for more developmental communications sponsored by native groups, universities, provincial and federal government departments and private communications carriers such as the Trans-Canada Telephone System. Direct TV broadcasting to individual homes was also made possible on a limited basis to 1.2 or 1.8 m TVRO earth stations connected directly to a TV set.

The present TELESAT system comprises ANIK C and ANIK D satellites which were manufactured by Spar Aerospace with Hughes Aircraft and

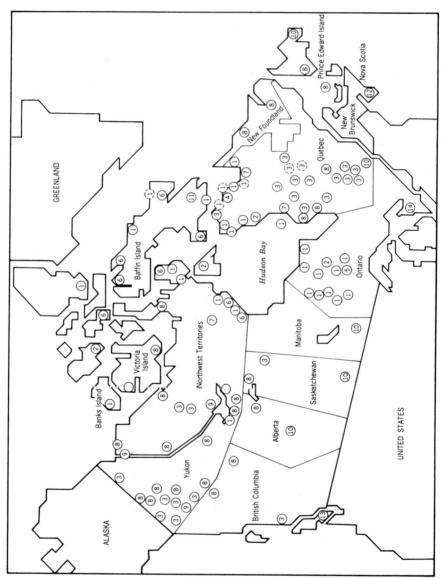

Figure 9.7. Canada's early ANIK satellite system.

204

launched during the years 1982 to 1985. Three ANIK C satellites have a design life of 10 years using only the Ku-band, whereas the ANIK D satellites have an 8-year design life and use only C-band frequencies. Canada's five orbital locations for these satellites are:

Degrees East	Satellite	Band	Launched	Main Uses
243	ANIK C-3	Ku	1982	General service
250	ANIK C-2	Ku	1983	Orbital spare
248	ANIK D-2	C	1984	Orbital spare
255	ANIK D-1	C	1982	Mainly TV

ANIK C satellites have 20 Ku-band transponders with 54 MHz bandwidth. ANIK D satellites have 24 C-band transponders with 36 MHz bandwidth. Major users of the satellites include Bell Canada, Canada

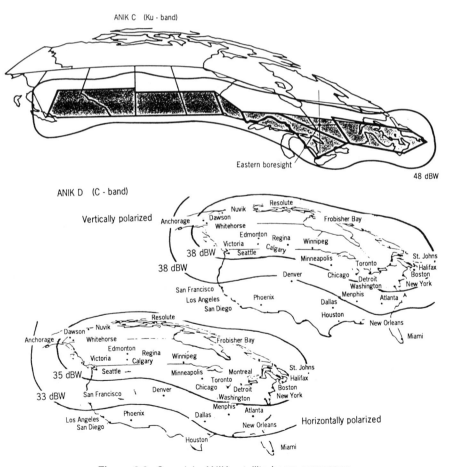

Figure 9.8. Canada's ANIK satellite beam coverages.

Broadcasting Corporation, Parliament, oil, music and sports services and also newspaper distribution printing services. Some C-band transponders are leased by U.S. companies for television distribution and transborder private communications services. However, there was excess capacity and TELE-SAT sold ANIK C-1 to a new far-east satellite company (Asiasat) in 1987.

TABLE 9.3   Canada's ANIK Transponder Applications (1986)

Location Degrees East	Satellite	Transponder Number	Frequency (MHz)	Service
242.5	Anik C-3	8	11.920	TV: Knowledge Channel
	(Ku-band)	9	11.961	TV: Access, Alberta
		10	11.967	TV: Premiere Choice
		12	12.048	TV: Superchannel
		17	11.730	TV: Atlantic
		24	11.939	TV: La Sette
		26	12.000	TV: Ontario
		32	12.183	TV: Radio Quebec
		10		Trunk Telephone and data services
247.5	Anik C-2 (Ku-band)			Spare in orbit
248.5	Anik D-2 (C-band)			Spare in orbit
251.5	Anik C-1 (Ku-band)			Spare in orbit Sold to Asiasat in 1987
255.5	Anik D-1	1		Spare
	(C-band)	2	3.740	TV: TSN Sports
		3	3.760	Spare
		4	3.780	TV: CTV/Global
		5	3.800	TV: CBC
		6	3.820	TV: Much Music
		7	3.840	TV: CBC
		8	3.860	TV: CHCH, Hamilton
		9	3.880	TV: CBC, Detroit
		10	3.900	TV: ABC, Detroit
		11	3.920	TV: CBS, English
		12	3.960	Faulty
		13	3.980	Faulty
		14	4.000	TV: TCTV
		15	4.020	TV: CBC, French
		16	4.040	TV: Parliament, French
		17	4.060	TV: CBC News
		18	4.080	TV: CITV, Edmonton
		19	4.100	TV: CBC, English
		20	4.120	TV: CBC, Montreal
		21	4.140	TV: PBS, Detroit
		22	4.160	TV: BCTV, Vancouver
		23	4.180	TV: NBC, Detroit
		24	4.200	TV: CBC, Parliament

Two ANIK E satellites were ordered for launch in 1990 from Spar Aerospace for US$180 million, which have more than three times the ANIK C Ku-band power, in order to expand the private business network capability. Two hybrid satellites are being built by Spar Aerospace for C$180 million which will contain twenty-four 36 MHz transponders at C-band, and sixteen 54 MHz transponders in the Ku-band.

The applications of the ANIK satellites in 1986 are shown in Table 9.3 and the satellite beam contours are shown in Figure 9.8.

CHILE

Chile, a long, narrow country stretching along the Pacific side of the southern Andes mountains, is most unsuited to terrestrial communications, although in earlier days, the major coastal cities were linked by telegraph cable. As well as the mainland, Easter Island in the central Pacific is also part of Chile and can only be reached by high frequency radio although satellite links started in 1987.

Chile started leasing transponder capacity from INTELSAT in 1977 using the minimum 9 MHz capacity to provide telephone communications between the capital, Santiago, and Punta Arenas in the extreme south of the country; the city of Coihaique was added in 1981 and the capacity leased was increased to 18 MHz. In 1984, an additional 36 MHz was leased so that television could be distributed nationwide, and satellite communications now provide telephone, data, and television services to the whole country including Easter Island. At the present time Chile is leasing 78 MHz global beam capacity from INTELSAT and has 23 earth stations mainly for television reception and redistribution.

CHINA

The People's Republic of China spent many years studying not only its own telecommunications needs for using satellites but also studying other countries' experiences. China, being one of the largest, most populous, and—by western standards of communications, least developed—is potentially one of the largest telecommunications markets in the world. Not only did many countries send delegations to China during the early 1980s in an attempt to enter this market, but China also sent many delegations to other countries to learn from other experiences. In fact, it was a great personal experience when I accompanied a nine-person delegation to INTELSAT in 1982 which included short visits to a number of U.S. satellite and earth station equipment suppliers. While China was doing these studies and researching others' experiences, they also started developing and launching an independent Chinese satellite, launching rocket, and earth station manufacturing facilities. China now has a wide range of capabilities of its own in the field of satellite communications.

As a result of all this preliminary work, China started using INTELSAT leased hemispheric transponders in 1985 as well as two similar transponders which were purchased for US$5.45 million in May 1986; these are expected to provide service until 1990. China had also been permitted some free use of a transponder under INTELSAT's Project Share during the preceding months so that educational experiments could be made within China and the most efficient use could be made of the new satellite facilities. In mid-1988 China had 40 earth stations as well as a television network.

The earlier experiments in rocket launching were also fruitful leading to the Long March launcher which is now capable of launching other countries' satellites and will be used by ASIASAT to launch its Hughes-built satellite in 1989.

China intends to launch its own satellite in 1989 and provided advance information to this effect in December 1985 for use of three orbital locations (87.5°, 98°, and 110.5°E). Three C-band satellites will provide telephone, telegraph, radio, broadcasting, and television services to the whole of China with approximately the same satellite power and characteristics as those presently used with the INTELSAT satellites. Meanwhile, the Chinese domestic satellite system is developing piece by piece in what seems to be a very unregulated manner with single earth stations, or groups of earth stations, being added for individual industries each seemingly reponsible for the development of its own private network—a philosophy which may be envied by some Western companies. This is all paving the way for China's own domestically manufactured, 16 C-band transponder satellite on its own launch vehicle in 1990–1991.

COLOMBIA

Colombia, like Chile, is predominantly mountainous and also has off-shore islands but is not quite so geographically separated. Colombia started leasing transponder capacity from INTELSAT in 1978 to provide voice communications between Bogota, Leticia, and the Caribbean island of San Andres. The leased network in 1988 consists of 45 MHz hemispheric beam capacity and 13 earth stations which provide voice communications and television distribution. Network planning assistance was given by INTEL-SAT under its Assistance and Development Program (IADP), and there are further plans to expand the television network to three channels.

Despite its early use of INTELSAT, Colombia has been a keen advocate of owning its own satellite and even solicited bids for the manufacture of its SATCOL satellites in 1982. In 1983 Colombia started the frequency registration process with the IFRB describing two C-band satellites colocated at 75°W. The satellite supplier bids were rejected and the plans for its own system seem to be dormant. The cost of national ownership to this comparatively small country would be enormously expensive compared to shared satellite facilities using INTELSAT. Other possibilities for the future could be an Andean regional satellite system that has been under consideration for many years, or using the U.S. PanAmsat satellite. The latter alternative, placing dependence upon a foreign private company, would not be politically attractive to a country that has been considering its own national satellite and it looks as though the use of INTELSAT will be the best solution for several years ahead unless PanAmSat can offer lower cost, equal reliability, and shared ownership.

DENMARK

Denmark, including Greenland, has been using INTELSAT domestic services since 1978 to provide voice and data communications to remote towns. Due to the transatlantic separation between the Danish mainland and Greenland, global beam capacity is needed, and 18 MHz was leased on INTELSAT's satellite at 307°E until a full 72 MHz transponder was purchased in 1988. There is one earth station, Blaavand, situated on the west coast of Denmark's mainland, one on an oil rig in the North Sea, and seven in Greenland.

The earth stations in Greenland are situated in Angmagssalik, Egedesminde, Godthaab, Qanaq, Scoresbysund, Umansk, and Upernavik.

FRANCE

France consists not only of the European mainland but also departments and overseas territories in the Caribbean, Indian, and Pacific oceans. Voice communications via satellite were introduced over the INTELSAT system at a very early stage on an individual channel leasing basis, but by 1976 traffic had grown sufficiently that France decided to lease half (18 MHz) of a global beam transponder to provide telephone or alternate television to the Indian Island of Reunion on a preemptible lease basis. In 1980, France took up a 36 MHz global transponder on a nonpreemptible basis to the Caribbean departments of Martinique, Guadaloupe, and Guiana. Both of the leases on INTELSAT were maintained until 1984 when France's own satellite was successfully launched, although experimental satellites had provided some services before that time. France has now two commercial domestic satellites operating or planned (TELECOM I and II) with a third as a ground spare.

TELECOM IB failed in orbit in 1988 and TELECOM IC was launched to replace it in 1988. In mid-1988 France continued to lease transponders from INTELSAT.

The French Government informed INTELSAT of its intentions to establish its own satellite communications network in 1979 to meet the growing demands; it was intended to be operational by 1983. The three TELECOM I satellites were ordered from MATRA Espace and based on the successful European OTS spacecraft platform with almost all parts of the satellite and

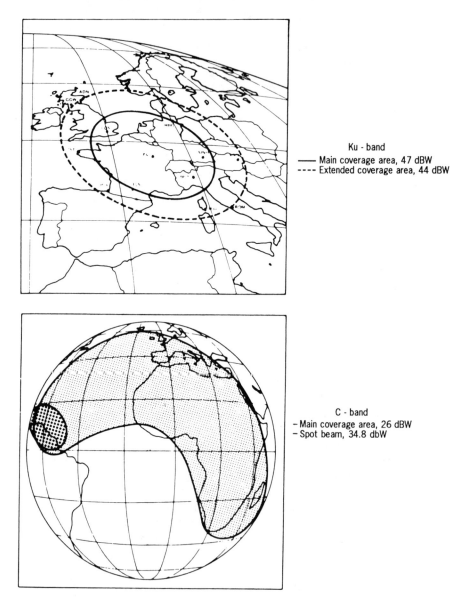

Ku - band
——— Main coverage area, 47 dBW
- - - - Extended coverage area, 44 dBW

C - band
– Main coverage area, 26 dBW
– Spot beam, 34.8 dbW

Figure 9.9. France's TELECOM satellite beam coverages.

all earth stations being of French design and manufacture. The satellites were designed to provide coverage for France and its overseas departments in the Caribbean and Indian oceans from orbital locations of 5° and 8°E. The coverage is shown in Figure 9.9. These satellites are also capable of providing other international and transatlantic services should international deregulation and competition with INTELSAT become possible.

The TELECOM satellite services are almost entirely digital, and the prime operator is France Cables et Radio which also operates most of the telecommunications in the overseas departments. Business services are also available between 2.4 kbps and 2.948 Mbps and earth stations can be installed on user premises.

Since 1977, France has also been planning its own DBS satellite, TDF-1, and this was launched in October 1988 to provide four high power television channels direct to French subscribers. TDF-1 was built together with a similar satellite for Germany by a joint French-German consortium named Eurosatellite at a total cost to France of about US$400 million. For this cost and only four TV channels, it is highly unlikely that services can be provided competitively with the medium powered satellites of INTELSAT, EUTELSAT, or Luxembourg's 16 channel satellite in 1988. However, the experimental experiences will have been very significant and France will undoubtedly be able to reap many benefits from its adventures into direct broadcast satellites.

GABON

Gabon asked for INTELSAT assistance in establishing a domestic satellite network early in 1985. The requirements were for voice, data, and television starting in 1986 and expanding later to three channels of digital television. Much preliminary planning had already been done with the help of a French consulting company (Bureau Yves Houssin); INTELSAT provided assistance under its Assistance and Development Program in devising the optimum transmission plan and also in providing a general training seminar in the capital city of Libreville in 1986.

When Gabon first approached INTELSAT, transponders could only be leased but the prospect of purchasing transponders was by then under consideration. The purchase of transponders became possible at the end of 1985 and Gabon decided to purchase transponder number 21, a 72 MHz east hemispheric beam transponder, of INTELSAT V F-2 located at 359°E. This satellite had been launched in December 1980 and the prorated purchase price was US$852,000 from October 1986 for the remainder of the satellite's life estimated as being until May 1989. The satellite could well be operational beyond 1989 and any extension would be to Gabon's advantage as no additional cost would be involved. Gabon was the first African country to purchase capacity from INTELSAT.

A further half transponder, adjacent transponder number 22, shared with Norway, was purchased later in 1986 for US$332,170. The combined capacity of 108 MHz is used to provide three television channels and 100 voice channels to a number of 7-M earth stations around the country.

The earth stations for the Gabon national system were supplied by Scientific Atlanta of the United States with a 15.2 m control earth station antenna manufactured by the Vertex Corporation. The control earth station will reportedly be equipped later for satellite, telemetry, tracking, and control facilities inferring that sometime in the future, Gabon is considering either its own satellite, or being a possible control station for a regional satellite system. Gabon has already stated its desire to purchase similar transponder capacity when the satellite is replaced.

The INTELSAT transponder coverage to Gabon is shown in Figure 9.10, and an EIRP of 28.9 dBW is provided to Gabon.

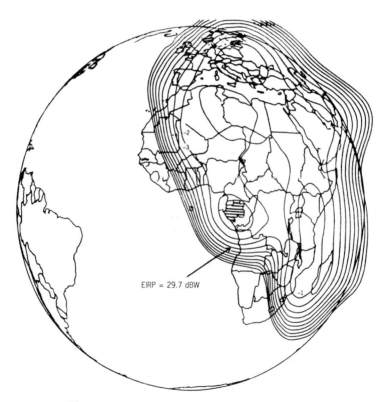

Figure 9.10. INTELSAT's satellite coverage for Gabon from 359° east.

GERMANY

Although Germany first started considering national satellite communications in the mid-1970s, political problems emerged and it was 1981 before plans were announced for using satellites to provide television services between West Berlin and the main area of West Germany. Five channels were envisioned and both INTELSAT and EUTELSAT were considered. Germany decided to lease three Ku-band transponders for five years from INTELSAT starting in December 1984, and also two transponders on EUTELSAT F-2.

Along with other European countries, Germany had also been planning its own national satellite, and had contracted in 1982 for the development and construction of a direct TV broadcasting satellite called TV-Sat-1 for launch in 1986 to be followed by a second satellite TV-Sat-2 in 1988. Launch problems during 1986 delayed the first launch until 1987. Although launched successfully on 20 November, TV-Sat-1 soon developed problems when its solar arrays would not deploy which blocked the receive antenna. After several months the satellite was considered a total loss. The next TV-Sat is scheduled for launch in 1989.

Both satellites were planned and are to be operated, along with all German telecommunications, by the Deutsche Bundesposte as part of the country's industrial rather than broadcasting policy. The TV-Sat satellites were part of a joint project with France's TDF satellites. Two other satellites called Kopernicus 1 and 2 are planned for 1988–1989 to provide all types of domestic services in the Ku-band and also to carry an experimental package at the higher Ka-band frequencies. The Bundesposte expects to use these satellites which will be combined with the domestic ISDN network, and will also introduce a new customer-dialled 64 kbps data service.

INDIA

India is yet a further example of a large underdeveloped country which early realized the enormous advantages that could be gained from developing a domestic satellite communications system. The major difference in the case of India was its determination to plan, develop, launch, manufacture, and operate as much as possible of the satellites rather than direct purchase as in the case of Indonesia and other countries. With this objective, India first coordinated and obtained approval from INTELSAT, for the INSAT satellite system in 1977 following successful experiments in making and using satellites for education and television services through NASA's experimental ATS satellite.

The government of India planned a multipurpose domestic satellite system, called INSAT, to provide telecommunications, television, broadcast and meteorological services. Twelve transponders, each with a bandwidth of 40 MHz in the C-band, were to be used for deriving heavy, medium, and thin route message and television program distribution services. The television broadcast services for direct reception by community sets were to be provided on two transponders operating with C-band up-links and L-band down-links. For the meteorological services, a radiometer operating in the visible and infrared bands would collect the required meteorological data for transmission to a central data processing earth station. There would also be a narrowband transponder in the 400/4000 MHz bands for collection of data from unattended land and ocean based platforms and their transmission to the central processing station. A matching ground segment with appropriate transmission and/or receiving facilities would be provided for the three types of services.

It was 1982 before the first INSAT satellite was launched and INTELSAT leased transponders were used to start implementing the earth segment from 1979 starting with 9 MHz, rising to 117 MHz in 1982, and then to 173 MHz in 1988. The reason for India continuing to lease from INTELSAT is the partial failure of the INSAT system.

The introduction of the INSAT system was hampered by launch problems. The first satellite was launched successfully in April 1982 on a NASA Delta launcher, but failed in orbit a few months later when its earth sensor locked on to the moon during an eclipse and all its station keeping fuel was used in trying to rectify the situation. The second launch, on a shuttle, was successful in 1983 but left India without a spare satellite. The replacement satellite, due to be launched by a shuttle in 1986 did not take place due to the shuttle launch disaster in January 1986. The INSAT satellite had no in-orbit spare and India continued to use INTELSAT leased capacity for support purposes. The third satellite was launched by Ariane in 1988, but this also failed in orbit shortly afterwards.

The first series of INSAT satellites used a new type of platform purchased from Ford Aerospace and at the time was one of the most complex commercial satellites yet built, being used for multi-use communications, quasi DBS television, and also meteorological observation purposes. The first two satellites cost US$130 million which included launch and a ground control station; the third replacement satellite cost US$58 million excluding launch all of which show the high cost risks involved in such national satellite systems.

The concept of the INSAT system using two satellites at 74° and 94°E is shown in Figure 9.11. The satellites were manufactured by Ford Aerospace to Indian specifications and requirements with a design life of seven years; thus, on account of the problems experienced, INSAT 1, launched in 1982, will have exceeded its design life before its backup satellite is launched.

For meteorological purposes, a very high resolution radiometer with visible and infrared channels with full earth coverage, and full-frame image every 30 minutes is used for round-the-clock weather observation. A

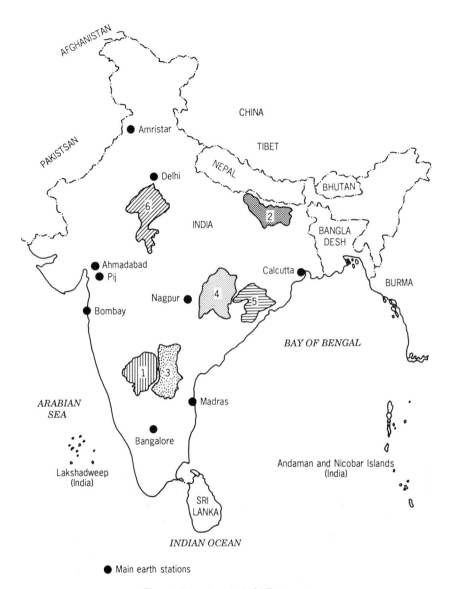

Figure 9.11. India's INSAT system.

weather data channel operates at 402.75/4038.1 MHz for relaying weather data from unattended platforms. The communications ground segment initially consisted of:

Five large earth stations with 11-m diameter antennas, 13 medium earth station with 7-m antennas

Eleven remote area stations with 4-m antennas

Four transportables with 4-m antennas

Two jeep transportables

One network control earth station

TV receive-only earth stations for direct reception

Seven of the larger earth stations came into service on the INTELSAT leased network in 1979, but the remainder were phased in after 1982 when the first satellite became operational.

The meteorological ground segment consists of:

A Meteorological Data Utilisation Centre (MDUC) at New Delhi for processing INSAT-1 Data received at the Delhi earth station by the P&T and transmitted to MDUC by microwave.

Secondary Data Utilisation Centres (SDUC) colocated at various forecasting offices of the Indian Meteorology Department which receive processed images from the MDUC by terrestrial lines

About 100 data collection platforms

A disaster warning facility

The MDUC has facilities for processing, analyzing, and storing the weather data. The data collection platforms access the satellite randomly using PDM-PSK modulation, and each platform can handle ten sensors for air temperature, wet bulb temperature, relative humidity, wind speed, wind direction, atmospheric pressure, platform housing temperature, rainfall, sunshine, and sea surface temperature.

The earth station networks using INSAT and INTELSAT have expanded despite the lack of spare satellite and a data network has been established using spread spectrum techniques and equipment supplied by Equatorial Communications. This network, called NICNET has its control center at New Delhi and the earth station and data controllers are being manufactured in India under license. NICNET is providing mainly government communications initially but is expected to expand quickly for public and private networks.

INDONESIA

Indonesia was the one of the earliest domestic satellite systems and resulted from lengthy marketing negotiations with a Washington D.C. consulting company and the Hughes Aircraft Company during the early 1970s. The first satellite was launched in 1976, the spare in 1977, and the system became operational in 1978. The satellite system was called Palapa which is now a word meaning "national unity".

National telecommunications are the responsibility of the Directorate

General of Posts and Telecommunications which is part of the country's Ministry of Communications. Three government-owned enterprises exist under this Directorate:

Perumtel   The national telecommunications administration
INDOSAT   The international services administration
PTINTI   A manufacturing organization

It was Perumtel who became responsible for the domestic satellite system.

Geographically, Indonesia was a leading candidate for satellite communications, consisting of some 5000 islands scattered quite widely in the south-west Pacific. It had only 250,000 telephones in January 1976 and the cost of any modern upgrade not using satellites would have been prohibitive. Using satellites, Indonesia expected to increase the number of phones to 650,000 in two years. By 1981, there were 500,000 lines installed showing that growth was less than expected but nevertheless very encouraging. Actual traffic growth is not generally available, but the two satellite transponder planning arrangements as given by a UNESCO report (PG1-84/WS/10) in 1984 were:

Transponder	Use
1	166 Demand-assigned telephony
2	Backup for transponder 1
3	212 Telephone circuits
4	600 Leased telephone circuits (MANKAM)
5	336 Telephone circuits
6	480 Telephone circuits
7	Indonesia Television (TVRI)
8	Leased to Philippines
9	Leased to Thailand
10	Leased to Malaysia
11	Spare
12	Occasional TV for ASEAN television

The first satellites, Palapa A, were the HS-33D type similar in many ways to the INTELSAT IV except for polarization and power arrangements which copied the United States domestic satellite design concentrating the satellite power into smaller beams for the use of smaller and less costly earth stations. The satellites had 12 transponders with a shaped beam antenna that could provide a down-link power of between 31 and 34 dBW to Indonesia and neighboring countries as shown in Figure 9.12. The terrestrial system included some 40 earth stations, 1200 installed voice circuits and national television. After the system had been working for about a year it was shown that traffic demand was considerably higher than had been expected and

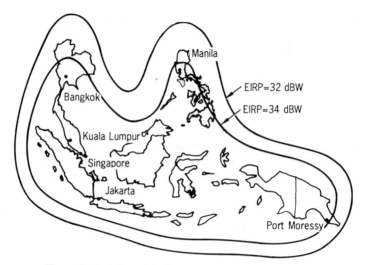

Figure 9.12. Indonesia's Palapa satellite beam coverage.

also that three transponders on the spare satellite had been leased to other countries at a rate of US$693,000 per year per transponder.

Although there had been early criticism of the project within and outside Indonesia, it did permit considerable improvements in the country's communications network. The first satellite, launched in 1976, had a seven year design life and Indonesia decided to replace them with two new satellites around 1983. Perumtel again contracted with Hughes Aircraft to supply two Palapa B satellites for US$75 million each plus launch costs of US$18 million. Palapa B-1 was launched successfully in June 1983 but the Palapa B-2 when launched in February 1984 did not reach its proper orbit for which Indonesia claimed and received US$75 million for its insured value. The insurers, Lloyds of London, in conjunction with NASA successfully retrieved the satellite from space in 1984 and also a WESTAR satellite that had experienced a similar fate. The Palapa B-2 satellite was subsequently resold by the insurers to a United States company and will be resold to Indonesia for a later relaunch.

Indonesia arranged to buy a replacement satellite (Palapa B-3) and arranged a further shuttle launch for 1986; it was eventually launched in 1987 on a Delta rocket by NASA.

It is very debatable whether the Palapa system has been a failure or a success. Two things are fairly certain; the national satellite system has been extremely costly for a lesser developed country, and the system has been very much underused. On the other hand, the system has greatly improved the country's communications and overall standard of living. Whether the results were worth the high cost is, again, debatable. It is also difficult to obtain figures for the costs and benefits, but the same benefits could probably have been derived at much less cost had the satellites been shared with other countries either regionally or via INTELSAT.

IRAN

Iran first stated its intention to own and operate its own satellite system in the late 1970s during the Shah regime, before the major political upheavals started, and also during the days of the oil boom. The plans were extensive, including three Ku-band satellites, known as Zohreh, at 26°, 34°, and 47°E. Each satellite was to have 12 Ku-band transponders and five higher power transponders for broadcasting purposes. These plans collapsed in the early 1980s and alternate means using INTELSAT were explored.

Iran first started showing serious interest in leasing capacity from INTEL-SAT in 1982 and began using the services of the Assistance and Development Program (IADP) to assist planning the refurbishment of existing earth stations which were hampered at the time by U.S. trade sanctions, and also in planning a domestic network. Requirements were separated into television and telephone requirements which could, if necessary, be on different satellites. The TV network would provide programs to over 100 town and villages using TVROs of about 2 m diameter.

Geographically, Iran is unsuited to using C-band on INTELSAT V because during the satellite design stages of INTELSAT, Iran had been placed between the main east and west hemispheric beams. At C-band, the only solutions were to use a zone beam transponder from the satellite at 66°E, a hemispheric beam from a satellite at 359°E, or a global beam transponder from 60°, 66°, or 359°E. None of these provided Iran with the best geographical advantage for using small earth stations. Unfortunately, the availability and plans for the Ku-band transponders were not known at that time. Iran nevertheless requested INTELSAT to lease two Ku-band transponders from 1985 onwards.

During 1984 and 1985 the availability of the Ku-band transponders became more certain, and also it became possible to purchase Ku-band transponders from INTELSAT much more economically, to Iran, than leasing them. Iran decided to cancel the leasing plans and instead purchased two Ku-band transponders on the INTELSAT F-2 satellite at 63°E in September 1986. This was followed by the purchase of three more Ku-band transponders on INTELSAT V, F-7 at 66°E in December 1986 to take effect in July 1987.

Iran has a total of five Ku-band transponders for its own domestic use. The two transponders on INTELSAT V F-2 cost just over US$4 million and the satellites have an expected design life up to the end of 1989. The three transponders in INTELSAT V F-7 cost about US$7.4 million and can be expected to provide service until the end of 1990. The intervening years will provide time for Iran to decide whether to continue using INTELSAT after 1990, or to pursue its own national satellite plans. Considering that the space segment for the first five years will have cost under US$12 million compared to an approximate US$300–500 million for owning its own, the economic answer would seem simple.

The initial earth station network for introducing a rural telephone system

Figure 9.13. Iran's domestic satellite telephone network.

all over the country is shown in Figure 9.13; this will augment and expand the existing terrestrial network using cables and microwave. As the earth stations are installed it is likely that the network will expand further as Iran has no shortage of transponder capacity during the early years.

During the war with Iraq, the Iranian international earth station complex near Asadabad was attacked at least twice. Iran is, naturally, not eager to release details of its earth stations or its future plans in the same way as many other countries. In addition, on account of various trading problems with a number of countries, Iran meets many problems in finding suitable earth station and equipment suppliers. For this reason, much of Iran's future satellite communications development is likely to depend on its ability to manufacture the ground equipment locally. Iran still plans to expand its use of satellite communications and is again considering launching its own satellite in 1992.

IRELAND

A fairly small island such as Ireland does not seem to be a likely candidate for owning and operating its own satellite but, much like the shipping industry, it provides a suitable "flag of convenience" from which to launch television services to the profitable British and European viewers from the orbital location at 31°W which it shares with the United Kingdom. Early in 1988 an Irish businessman, James Stafford, chairman of a new company Atlantic Satellites, was chosen by the Irish government to start a DBS television service. Atlantic Satellites was formed as a joint venture 20% owned by James Stafford, and 80% owned by the Hughes Aircraft Company. Two DBS satellites were planned to be operational early in 1989 with one additional ground spare at a total cost of about US$400 million. Due to the industry's problems in 1986, these plans were postponed until the early 1990's. Each satellite will have four high power transponders for DBS services to private homes, and eight lower powered transponders for European cable TV distribution.

The satellites will also be technically capable of providing transatlantic services if the INTELSAT cooperative monopoly is broken. Although Atlantic Satellites has not stated its intention to compete with INTELSAT, the technical capability exists should the market open up and become attractive.

The two-satellite system will provide eight 120 W transponders and sixteen 50 W transponders controlled by an earth station at Shannon. In exchange for the Irish Government license, a 2.5% revenue fee will be paid to the Irish Department of Communications.

ISRAEL

An African-Mediterranean System (AMS) satellite was under consideration by Israel for some years as a response to the ARABSAT regional system. AMS was developed as a private company using European finance and American technology. Coordination with INTELSAT caused problems mainly due to active opposition by the Arab members of INTELSAT, and no decision was made by the INTELSAT Board of Governors with the result that the project is now stalled indefinitely.

Israel also asked INTELSAT if it could lease two Ku-band transponders starting in 1984, and later changed this to three Ku-band transponders starting in 1986, a lease agreement for which was approved by the Board in March 1984. When it became possible to purchase transponders late in 1985, Israel was third in line and purchased three east spot beam Ku-band transponders on INTELSAT V F-2 at 359°E for a total of US$4.32 million. This satellite, launched in 1980, had a life expectancy up to May 1989 after which Israel intends to purchase three transponders on the replacement satellite. The spot beams provide perfect geographic coverage of Israel as shown in Figure 9.14.

Israel's purchase of three transponders for the years 1986–1989 for US$4.32 million was a great financial saving to the country as the planned domestic noncancellable lease of the same transponders for the same period would have cost US$5.95 per year and a total of US$23.82 million.

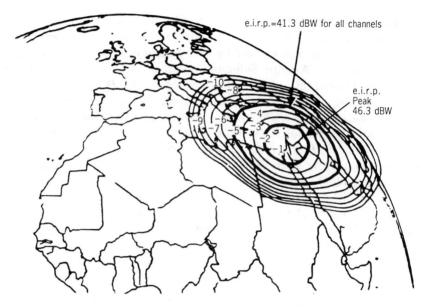

Figure 9.14. Israel's Ku-band coverage from INTELSAT.

ITALY

Italy has been involved in the space industry for many years and like some other European countries is using both INTELSAT and EUTELSAT for its domestic satellite services before launching its own commercial national satellite, ITALSAT, around 1990.

In 1982 Italy requested INTELSAT to lease a full 72 MHz Ku-band transponder for five years between 1985 and 1990, and also a 9 MHz segment of a Ku-band transponder for a similar period. Italy actually started leasing a 72 MHz hemispheric beam transponder in June 1985 but when INTELSAT introduced the sale of transponders at the end of 1985, Italy purchased a hemispheric transponder instead on a satellite located at 359°E. This transponder is expected to provide good service until early 1990 and was purchased for US$2.586 million. The preemptible lease for a similar transponder costing US$1,360,000 per year was cancelled but two additional Ku-band spot beam transponders were purchased in 1987. Italy's own national satellite is expected to be launched by Ariane 4 in 1990 or 1991 about three years after the original planned date. The single satellite is being manufactured by the Italian manufacturer Selenia Spazio for about US$230 million and will contain Ka-, Q-, and V-band transponders to experiment with digital services at these frequencies. Six Ka-band transponders will provide up to 11,000 voice circuits and three will provide data services at Q- and V-band frequencies. The experience for ITALSAT has been gained from an earlier experimental satellite, SIRIO. This was sponsored by the Italian National Space Plan which had been started in 1978 and gave full expected service until 1985 when its remaining propellant was used to move it to its final resting place at 75°E.

ITALSAT was originally intended for experimental use for interfacing the terrestrial network, providing emergency backup, and introducing new all digital services. If successful, it will be used commercially to augment the existing terrestrial network and add flexibility for new services. About 15 earth stations are likely to be used to link with the terrestrial network, but this number could well be much greater if the system proves satisfactory. Although television transmissions are not planned it seems more than probable that these plans will change as television in Europe at the present is about the most lucrative use for satellites, and the Ka-band experiments are successful.

One of the unique applications for satellite communications used by Italy is its ARGO network for emergency telecommunications. This network was developed jointly by Telespazio and Selenia Spazio for earthquake surveillance, public safety, and portable earth stations that can be carried by helicopter to disaster areas. Each portable earth station has a 2.4 m antenna for relay communication and transmitting television. There are also over 100 small unattended terminals spread around the country. The ARGO network uses a EUTELSAT Ku-band transponder.

IVORY COAST (COTE D'IVOIRE)

Ivory Coast constructed two earth stations for domestic use in 1987 and started leasing 9 MHz capacity from INTELSAT to link the old capital, Abidjan, with the new capital city of Yamoussoukro. The Ivory Coast is a comparatively small country on the west coast of Africa and already has one of the more advanced internal telecommunications in that part of the world. The new satellite link will form the first part of a larger satellite communications system within the country.

JAPAN

For its international telecommunications purposes, Japan's signatory to INTELSAT and INMARSAT has always been, and remains, the Kokusai Denshin Denwa Company (KDD). For its domestic telecommunications, the Nippon Telephone and Telegraph Company (NTT) held a monopoly until deregulatory changes took place in 1985. For its other space activities, a space activities commission was formed in 1968 to advise the Prime Minister on space policy and to coordinate Japanese industrial and research activities. These activities include communications, broadcasting, meteorological, navigation, geodesy, and remote sensing; the practical applications became the responsibility of a new National Space Development Agency (NASDA) in 1969. NASDA is also responsible for tracking and the space management of all Japanese satellites.

Numerous experimental satellites were launched starting in 1970 and satellites with practical applications were launched starting in 1975. Most of these satellite launches were performed using Japan's own launch vehicles and facilities, the first launch being on an "L" rocket in 1970. The launch facilities are provided at two space centers, NASDA's Tenegashima Space Center and the University of Tokyo's Kagooshima Space Center.

Following the deregulatory law changes of 1985, there are now three general domestic telecommunications satellite companies in Japan.

1. The government-owned Nippon Telephone and Telegraph (NTT) has two Japanese CS satellites that are small, heavy, and costly. These satellites were developed as prototypes for Japanese industry and also the higher Ka-band of frequencies, which may prove very rewarding to Japanese industry during the next decade. There has, however, been much criticism of these satellites and their cost, especially by NTT who are the principle users.

2. The Japan Space Communications Corporation (JSCC) is a group parented by the Mitsubishi group and is procuring two Superbird satellites manufactured by Ford Aerospace.

3. The Japan Communications Satellite Company (JCSAT) is a consortium of three Japanese companies (NEC 20%, Mitsui 35%, and C. Itoh 35%) and the United States' Hughes Aircraft Company (10%). JCSAT ordered two Hughes satellites, the first of which was launched in March 1989 by Ariane.

All three companies will provide a full range of business services as well as television. When all three companies' satellites are in orbit, Japan will have about 150 transponders for its own domestic use, and it is difficult to imagine how these could all be used domestically. With this excess capacity Japan could become an extremely competitive supplier of satellite services for Far Eastern countries. These satellites are summarized in Table 9.4.

The DBS broadcasting satellites are owned and operated by Nippon Hoso Kyokai (NHK) which operates two television and three radio networks. The first experimental broadcast satellite (BSE) was launched in 1976 and was replaced by BS-2 satellites in 1984. All of the BS-2A satellite's three power amplifiers failed after four months which was a severe disappointment as BS-2A would have been the world's first operational DBS TV satellite but, like the CS satellites, the research and development are likely to be rewarded during the next decade. BS-2B was launched in February 1986 and has been providing two channels of DBS since July 1987. Although in a much more limited fashion than had been hoped, it is providing the world's first DBS service and half a million households were equipped to receive its programs by the end of 1987. The small (45 cm) antenna and associated tuning equipment cost less than US$1000 (135,000

TABLE 9.4 Summary of Japan's Domestic Satellites

Orbit Location (°E)	Satellite	Launch Year	Purpose
110	BS-2B	1986	Broadcasting
132	CS-2A	1983	NTT General Services
135	CS-2B	1983	NTT General Services
140	GMS-3	1984	Meteorological
150	JCSAT-1	1988	JCSAT General Services
154	JCSAT-2	1988	JCSAT General Services
160	Superbird-1	1988	JSCC General Services
162	Superbird-2	1989	JSCC General Services

Yen). High definition TV experiments started late in 1987 and full service is expected by 1990 which will further enhance Japan's domestic and international capabilities in the field of satellite telecommunications.

KOREA

The Republic of Korea has been exploring the possibilities of a domestic satellite network since at least 1982, and the idea was emphasized during a World Communications Year Conference held in Seoul during August 1983.

Geographically, Korea is small compared to many countries being only 38,000 square miles (about the same as Guatamala) but quite densely populated with about 36 million people. Compared to many countries it is well developed industrially, commercially, and politically. It is certainly a potential user of domestic satellite communications, but with the aid of outside influences the requirements became extremely large and included needs for television distribution networks, mobile applications, customer premise facilities, all having a forecasted need for about twenty 36 MHz transponders.

Responsibility for planning the Korean domestic system was given to the Korea Electrotechnology and Telecommunications Reseach Institute (KETRI) which issued a Request for Proposal in 1982 for consulting companies to assist it. The services and functional objectives of the Korean Satellite System were defined as being:

1. Direct broadcasting   3 Transponders
2. Video relay   2 Transponders
3. Communications
   - 10 Major city telephony   1000 channels
   - 100 Data networks   2000 channels
   - 20 Remote island stations   200 channels
   - 50 Mobile transportables   800 channels
4. Regional TV relay during the 1988 Olympic Games

INTELSAT offered the services of its Assistance and Development Program throughout the planning and implementation process, but suggested a gradual development approach as the proposed requirements were not likely to be introduced in a short space of time. INTELSAT suggested a Ku-band leasing approach starting in 1984–1985 for up to three Ku-band spot beam transponders on a Pacific ocean satellite which would provide a powerful EIRP of 48 dBW to Korea at an annual cost of US$5.76 million for all three transponders. Over a five-year period between 1987 and 1992, it was estimated that the net space segment cost would be about US$15 million taking the INTELSAT financial arrangements into account and at the 1983 charging rates.

The types of services possible from these transponders would have been two channels of quasi-DBS and four TV distribution channels with one or two of these being used for general services until such time that a Korean satellite was warranted commercially.

Other consultants provided different viewpoints which proposed the need for a Korean-owned satellite system to meet their needs and which would be much less costly than leasing from INTELSAT, but as in many cases of this nature, the figures presented were based upon the satellite being fully loaded and commercially viable from the outset which is an impossible supposition in most cases. One cost estimate provided to Korea was that for a seven-year period it would cost US$300 million to use INTELSAT but only US$160 million using a Korean satellite.

The Korean Government decided to postpone the project, at least for several years, and as yet no firm commitments have been announced for expanding Korea's domestic satellite services. Apart from a dedicated Korean national satellite and using INTELSAT, there are now a number of potential satellite operating companies emerging who could offer to supply the services to Korea.

LIBYA

Libya, geographically similar to its neighbor Algeria, also leases a 36 MHz transporter from INTELSAT and has been doing so since 1981. Originally a global transponder, the present transponder leased is an east hemispheric beam transponder on one of the Atlantic region satellites which is used in a similar fashion as several other countries provide one television channel and up to about 400 SCPC telephone channels. The 13 earth stations in different parts of the country use this transponder; they have been working since the original earth station network was supplied by NEC of Japan in 1981.

LUXEMBOURG

The Grand Duchy of Luxembourg with an area of less than 1000 square miles, and smaller than the United States' smallest state of Rhode Island would, at first glance, be an unlikely nation to need and launch its own national satellite. However, for many years Radio Luxembourg has covered Europe with one of its few popular, light, and commercial radio programs using the long wave. Therefore, it is not quite so surprising that in the era of television, Luxembourg should plan its own Europe-wide television broadcasting satellite which has been called GDL, LUXAT, SES, or ASTRA by its operator-Société Européene des Satellites (SES).

Early in the 1980s, a DBS project was considered but found not viable. In 1983 following a visit from Clay Whitehead (of the U.S. National Exchange satellite company, and former head of the Office of Telecommunications Policy under President Nixon), the project was revised to a quasi-DBS system for distributing cable TV channels using U.S. satellites. The Luxembourg government was supportive of the project and two RCA 4000 satellites were ordered for launch in 1986 and 1987; this was postponed about a year due to commercial problems in Luxembourg, and the whole project was initially expected to cost about US$200 million.

The ASTRA satellite system needed to be coordinated both with INTELSAT and EUTELSAT, but it was EUTELSAT who had most to lose by the appearance of commercial rival broadcasting television programs from the heart of Europe. Although there were no technical problems, there were severe economic side effects as EUTELSAT signatories (of which Luxembourg is one) are obliged not to support other systems that do economic harm or threaten the viability of EUTELSAT; ASTRA will do just that. To a somewhat lesser extent, ASTRA will also pose economic harm to INTELSAT as it could take away some of the cable TV networks presently

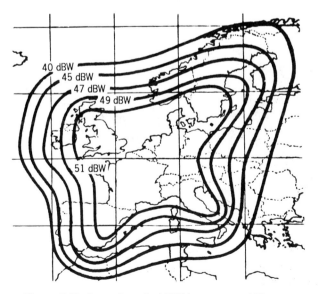

Figure 9.15. Luxembourg's ASTRA coverage of Europe.

distributed by INTELSAT. The systems most likely to be adversely affected are the DBS systems of France and Germany as the ASTRA prices are likely to be a mere fraction of the DBS costs.

The coordination process started with EUTELSAT in 1986 but no accord was reached until August 1987 when EUTELSAT recognized ASTRA and coordination was completed. Eleven of the transponders are to be leased to British Telecom.

Each of the two ASTRA satellites will have 16 operational medium power 45 W transponders beaming to small antennas of less than 1 m diameter which are expected to cost less than US$600 each. This type of transmission will compete directly with France and Germany's DBS satellites which require smaller antennas but which will cost a lot more for the space segment use. ASTRA will also cover a much wider area of Europe.

European TV and cable companies such as Scandinavia's SCANSAT, Thames Television, the British Broadcasting Corporation, Premiere, and the United States' MTV have shown interest, and the outcome of the venture will be interesting to watch in the years ahead. The coverage pattern of ASTRA is shown in Figure 9.15.

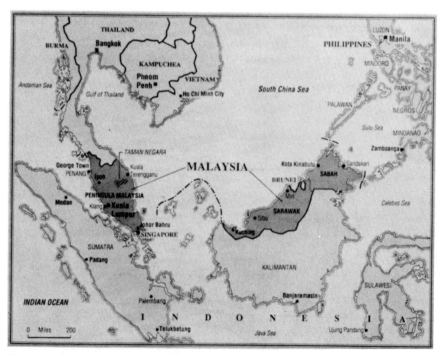

Figure 9.16. Map of Malaysia and its separated parts (Copyright © 1987 The Economist Newspaper Ltd. All rights reserved).

MALAYSIA

Malaysia is a country of two large parts separated by between 500 and 1000 miles of the South China Sea (Figure 9.16), and was one of the first countries to take a transponder lease from INTELSAT in 1975 which provided telephone and television services between the two areas. There were two earth stations, one near the capital city of Kuala Lumpur, and one near Kota Kinabalu in Eastern Malaysia.

At the end of the five year lease, service was taken from INTELSAT and transferred to Indonesia's PALAPA satellite for five years until 1984 when the service was transferred back to INTELSAT.

MEXICO

Mexico flourished during the oil boom in the 1970s and telecommunications development was a necessary part of this boom. Mexico first asked INTEL-SAT In 1979 to start providing domestic services for distributing four television channels during 1980 and 1981. Shortly afterwards the government requested a number of companies to perform feasibility studies for a national satellite system and finally went to international tender for a two satellite system in December 1981 for delivery in 1985.

The Request for Purchase (RFP) was simple and included:

Twenty-four 36 MHz transponders at C-band

Frequency reuse using linear polarization

An EIRP higher than 36 dBW

A design life of 10 years

It was clear that the Mexican approach to the need for its own domestic satellite system was political rather than of economic use or necessity. With no detailed requirements, Mexico first asked INTELSAT to lease four INTELSAT IV spot beam transponders starting in 1981. In mid-1981 this was revised to six transponders from 1982 to 1984, and for 40 transponders from 1985 onwards. Later in the same year (1981) Mexico issued its RFP and in 1982 contracted with the Hughes Aircraft Company for two satellites to be launched three years later in 1985 for a total cost of US$150 million funded by a loan guaranteed by the United States Export-Import Bank.

Both satellites were launched successfully and are called MORELOS after Jose Maria de Morelos y Pavon, a hero of the Mexican revolution. The satellites are located at 113.5° amd 116.5°W, and each satellite has both C- and Ku-band transponders. At C-band there are six 72 MHz transponders plus two spare, and at C-band there are twelve 36 MHz transponders plus two spare which are used for general telephony. At Ku-band there are four

108 MHz transponders plus two spare that are available for television distribution. The amount of traffic carried on the system is not known, and there is generally believed to be a lot of spare transponder capacity; the spare satellite in orbit is reportedly completely unused.

The satellite beam contours of MORELOS, like most national satellites, only provide coverage of Mexico and the immediate geographic area around it. At C-band MORELOS provides an EIRP of about 36 dbW to most of the country falling to 33 dBW at the edges of the beam; at Ku-band it provides about 45 dbW falling to 40 dBW at the beam edge.

Prior to the launch of MORELOS, the earth stations were used for receiving television over the leased INTELSAT transponders. These earth stations had 11 m antennas in the large cities, 7 m antennas in about 200 smaller towns, and plans for up to 1000 5-m antenna earth stations in villages. All the earth stations were of U.S. (Scientific Atlanta) design, but were mainly manufactured in Mexico under the name of DIGISAT. They were designed with linear polarization for use with a U.S. domestic type satellite and not for use with INTELSAT satellite, which all have circular polarization at C-band. Consequently, they needed to receive the maximum power from INTELSAT satellites and as a few of them were not equipped with even elementary tracking systems, several did not operate satisfactorily with INTELSAT for these reasons.

Shortly after the launch of the first MORELOS satellite, the great earthquake hit Mexico City in September 1985 (see also Chapter 4).

MOROCCO

Morocco started leasing a 36 MHz global or hemispheric beam transponder from INTELSAT in March 1982 for interconnecting the capital, Rabat, with Dakhis on the south west coast and Laayorne near the border with Mauritania and the disputed western Sahara region.

MOZAMBIQUE

Mozambique, a comparatively large but very undeveloped coastal country in southeastern Africa has been considering using INTELSAT for developing its domestic telecommunications at least since 1981, but its internal woes kept hindering the decision process and also in obtaining funds. The earth stations had been completed when Mozambique finally asked INTELSAT to lease the smallest 9 MHz capacity in 1985.

The country's international earth station is at Boane near the capital city, Maputo, in the south. The main domestic earth station is on the same site, and communicates with two other earth stations at the principal port city of Beira in the center of the country, and Nampula in the north.

The 9 MHz is expandable but at least it permits basic, high quality phone and telex services between the major cities without relying on a deteriorating terrestrial system which is often interrupted by terrorist activities.

NEW ZEALAND

The two main islands comprising New Zealand surprisingly waited many years before deciding to use satellites for domestic purposes—probably because there already existed an excellent terrestrial system. During the early 1980s New Zealand did, however, start considering using INTELSAT for television distribution as well as a backup for terrestrial networks, and eventually decided to lease 18 MHz from INTELSAT starting late in 1986 with plans to expand later if operation of the first lease turns out to be successful. The initial intention was to use the lease only for 8 Megabit digital transmission between the cities of Wellington on the North Island and Christ Church on the South Island.

As well as the lease from INTELSAT, New Zealand also decided to lease on 30 W transponder and half a 10 W transponder on Australia's third AUSSAT satellite which was launched in September 1987. Australia is also considering providing specialised facilities on the next breed of satellites (AUSSAT II) in the early 1990s that could provide DBS television through four 54 MHz Ku-band transponders.

The Australian satellite system is turning out to be particularly successful and in the long term, in view of the close relationship between the two countries, it seems natural that New Zealand will make more use of AUSSAT. However, INTELSAT's services to the region will become much more attractive when the INTELSAT VII series is introduced in 1992, and use of the international system may be preferred.

NIGER

This central Saharan desert country of almost half a million square miles has only about 5.5 million people—only 11 people per square mile. Satellite communication was recognized as being the only means of connecting its few towns with a high quality telecommunications service, and Niger took a five year lease of a hemispheric beam transponder starting in August 1981. The transponder provides telephone, telex, and television services for four of the largest towns, Niamey (the capital), Agadez in the center, Diffa at the south eastern border with Nigeria, and Bilma in the north.

NIGERIA

The Nigerian domestic satellite system was established in 1976 and was one of the earliest users of INTELSAT for this purpose: it was immediately

INTELSAT's largest domestic service user as the country leased three full global transponders almost from the start. Nigeria is among the few African countries that have mounted aggressive campaigns to improve their telecommunications networks and this trend continues despite many problems.

The system began operations early in 1976 and was introduced in three phases to provide nationwide coverage of the World Black Festival of Art and culture, FESTA 1977, with 19 earth stations. The system was devised, designed, and supplied by the United States' Harris Corporation to provide three regional services within the country as well as nationwide television distribution. Each of the three global transponders leased from INTELSAT contained one television carrier and up to 400 single channel per carrier (SCPC) telephone channels. The whole network was operated and maintained by the internal Posts & Telecommunications Department of the Ministry of Communications as distinct from the Nigerian External Telecommunications Corporation (NET) that operated and maintained the country's earth stations for overseas telecommunications. This arrangement continued for nearly ten years which was rather unfortunate as it divided the country's limited engineering resources.

Although the network provided innovative and expanded television and phone services to major cities all round the country, the network seemed to be beset with many practical problems which deprived the country from reaping all the advantages that it should from such an advanced system. Troubles included lack of sufficient interfaces between the earth stations and the terrestrial telephone network, an unsuitable power supply system that necessitated the full-time use of diesel generators at each earth station, and a constant problem with obtaining fuel and spare parts. It is little comfort to know that similar installations in Sudan and Uganda suffered from similar problems. Despite the system's shortcomings, many advances were made and the government was sufficiently pleased that further studies were encouraged to expand the use of satellite communications within Nigeria. Between 1976 and 1985 the Nigerian Government spent a total of about US$25 million for a decade of leased satellite capacity which compares very favorably with the near US$100 million had Nigeria decided to procure and launch its own satellites in 1975. A second series would also have been needed before the mid-1980s costing another US$300 million or more.

A Nigerian national satellite system was first proposed in 1977. Preliminary studies performed by U.S. consultants suggested a two satellite configuration in 1982 at a total cost of US$300 million for the satellites only. A further study was performed in 1983 with the help of expertise from the ITU which recommended a new system using Ku-band satellites, replacing all the earth stations on a phased basis, for a total of US$450 million including low cost earth stations. This was followed by another study performed by a satellite manufacturer that suggested a Nigerian-owned satellite could provide regional coverage to other West African countries. This report promised many long-term advantages, both economic and political, if Nigeria pursued the path of its own satellites at a basic space segment cost of US$350 million.

One major problem in any study of this nature is the total lack of forecasted use of satellite facilities whether they be leased or purchased outright. Lack of such basic data permits any study to find its way toward any desired conclusion; one study noted that Nigeria's armed forces would need six full 36 MHz transponders, six transponders would be needed for a

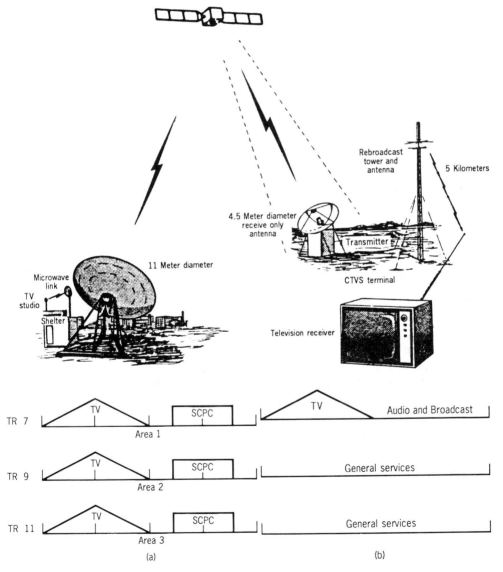

Figure 9.17. Nigeria's domestic satellite system.

national trunk network, and four others would be needed for television and rural links.

As an outsider it seemed clear from all these studies that a Nigeria-owned satellite system would be vast waste of already limited development funds, but proponents of the separate system continue to press the issue.

Meanwhile, although Nigerian terrestrial communications diversified and expanded, the need for satellite communications remains great, and many commercial interests are pressing to have modern business communications available to them which, given the inadequacies of local area networks, can only be provided through user-premise earth stations; these are now being installed in a number of key areas. These new services are to be supported by considerably increased efforts by the new Nigeria Telecommunications Ltd. (NITEL); that is, the unified telecommunications company replacing the previous internal and external separate administrations, to expand its training, maintenance, research, and monitoring services.

The present Nigerian satellite system, still using the three transponders leased from INTELSAT, is shown in Figure 9.17.

Further likely developments in Nigeria before 1990 could be:

Further deregulation
New and increased use of customer premise earth stations
Introduction of digital business services
Wider television and rural services
Growth using INTELSAT for domestic services

NORWAY

Norway first leased half a global beam (18 MHz) transponder from INTEL-SAT in 1975 for communications between the mainland and its North Sea oil fields. The leased capacity was expanded to 27 MHz in 1986 and later replaced by purchased transponders more economically.

The original leased system, which has changed little since 1976, connected the mainland to 8 m diameter antenna at earth stations located on the major oil fields of Ekofisk, Valhall, Frigg, Statfjord and a 13 m earth station at Isfjord on the Artic Ocean island of Spitzbergen which is 78°N. Because of the extreme northern location of the Isfjord station, and the problems of satellite visibility with an elevation angle of only 1.4°, the system was constrained to operate with INTELSAT satellites located at 342° or 359°E.

The services provided were voice, data, and facsimile. Earth stations were added later at Riser, Guilfaks, and Ula. The satellite services between the mainland and the oilfields are extended to production platforms by terrestrial radio links thus linking all oil rigs to the national terrestrial system or to the international system (Figure 9.18). The whole network was

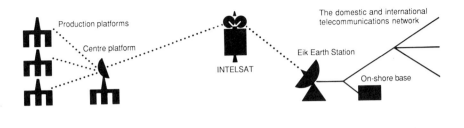

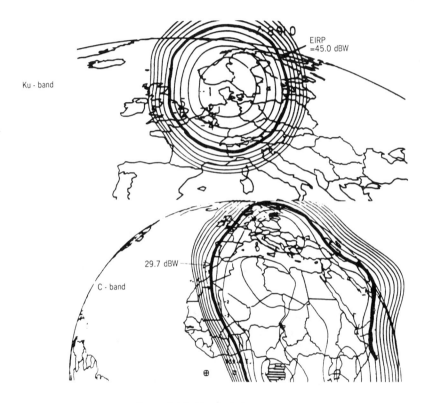

Figure 9.18. Norway's beam coverages.

transferred to half of a hemispheric beam transponder that Norway and Gabon purchased from INTELSAT in 1986.

Three west Ku-band spot beam transponders were also purchased by Norway from INTELSAT in 1986 for distributing the SVT-1, SVT-2, and New World television channels. All the purchased transponders are on INTELSAT V, F-2 located at 359°E, which was launched in December 1980 and is expected to provide full service until mid-1989. The total price for the three Ku-band transponders and half of a C-band transponder was under US$5 million. Another transponder is leased from EUTELSAT for distributing Norway's NRK television channel.

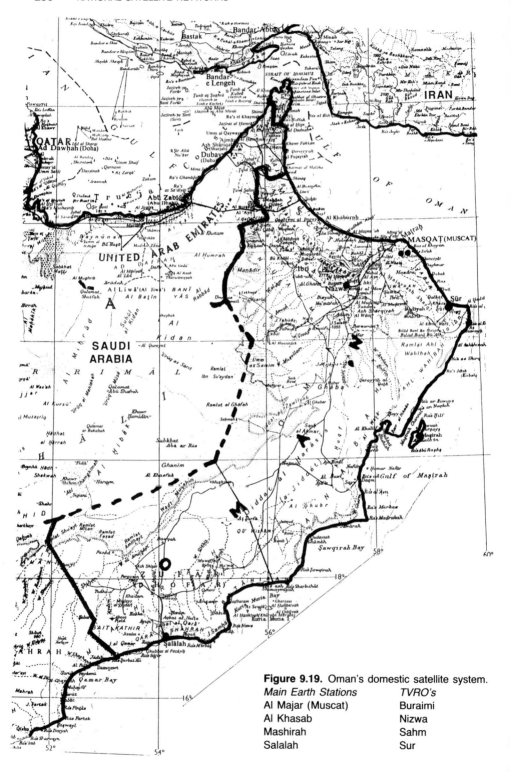

Figure 9.19. Oman's domestic satellite system.

Main Earth Stations	TVRO's
Al Majar (Muscat)	Buraimi
Al Khasab	Nizwa
Mashirah	Sahm
Salalah	Sur

OMAN

The Sultanate of Oman was another early user of INTELSAT's preemptible domestic lease service when it sought and obtained approval to lease a global transponer in 1977, which continued until November 1986 when the service was transferred to the ARABSAT satellite.

Geographically, Oman is of great strategic importance as its coastline covers much of the Arabian peninsula and all the way north to the narrow Strait of Hormuz, so guarding the eastern approaches to the Arabian or Iranian Gulf and the world's biggest oil trading route (Figure 9.19).

The single INTELSAT global transponder provided one national television channel and capacity of up to 400 SCPC voice channels between the four main earth stations near Muscat (the capital), Al Khasab on the Strait of Hormuz, and the coastal towns of Masirah and Salalah. The 400 voice channel capability seems plenty for years to come, but could be filled quickly if more earth stations are built in the remoter areas of the country.

PERU

Peru has three different geographic areas—coastal plain, the Andes mountains, and the Amazonian jungle—and is ideally suited for satellite communications which offer the only means of providing a communications network. On account of these terrestrial barriers, as recently as 25 years ago, telecommunications were limited internationally to an old telegraph cable running up the coast from Chile linking it with the British Cable and Wireless Group network, and high frequency connections provided by other nonPeruvian companies such as the West Coast of America Telegraph Company (Cable & Wireless), All America (RCA), and Western Union. Nationally, telecommunications were limited to the major cities that were linked mainly by radio or standard telephone wires.

After the formation of a national telecommunications company, ENTEL, international telecommunications by satellite were introduced through an earth station at Lurin, near Lima, in July 1969. The Peruvian national communications network consisted of a trunk radio relay network supplemented by a secondary rural network, until the domestic satellite communications system started in 1978 leasing 9 MHz from INTELSAT.

While considering all the options available in both satellite availability and the earth station manufacturer's market, ENTEL decided to make the maximum use of its own engineering expertise, expand it where necessary, and further train its own staff to take care of the network. With regard to the satellites available, this was limited to INTELSAT although an Andean Regional system was being considered. ENTEL has consistently kept track of the availability of other satellite systems, including those of the Brazilian satellite and the United States' PANAMSAT satellite. So far Peru remains with INTELSAT from which it has purchased two hemispheric beam transponders, but one transponder has also been purchased from PanAmSat.

The first stage of system development was building a second smaller earth station alongside the international earth station at Lurin, and to introduce telephone service with the towns of Iquitos on the Amazon River, Tarapoto high up in the Andes mountains and Pucallpa. These 13 m earth stations were supplied by NEC of Japan. Lurin was equipped with 100 SCPC channels, Iquitos 56, and Tarapoto and Pucallpa each were equipped with 28 channels. The earth station at Pucallpa was later moved to Quillabamba.

The second stage of Peru's domestic satellite network was completed in June 1981 and included installation of three more earth stations manufactured by Philips in the towns of Chachapoyas, Caballococha, and Contamina in the areas of Amazonas, Loreto and Ucayali having 12 m earth staions and a capacity of eight SCPC channels as well as television reception.

Two more stations were installed during the second half of 1981. The eighth station was at the agricultural and cattle rearing center at Chavez Valdavia in the Amazonas district. This was an 8.5 m earth station supplied by the United States' COMSAT General which had a capacity of four SCPC telephone channels. The ninth earth station, again supplied by Philips, had a 10 m antenna belonging to the Occidental Petroleum Company and was equipped with four SCPC telephone channels and television reception.

Three more earth stations were installed during 1983 at Juanjui, Saraposa, and Tocache Nuevo all in the San Martin district. These were part of an experimental program which ENTEL conducted with the cooperation of the U.S. Agency for International Development (AID) called the Rural Communications Services Pilot Project, ENTEL-AID. These earth stations were all 6.1 m supplied by the Harris Corporation, each with a capacity of four SCPC channels. The towns of Bella Vista, Huicungo, Pachiza, and Tingo de Saposoa were connected to the Juanjui station by VHF/UHF radio relay links. Peru's earth station network development is shown in Table 9.5.

Television was introduced in August 1982 using an additional 36 MHz lease from INTELSAT, and a further 9 MHz lease was started in August 1984 to accommodate traffic increase requirements. Peru plans to expand the television distribution network in 1988 to a total of 82 earth stations around the country.

The primary purpose of the experimental project was to demonstrate how satellite communications can help the development of rural areas, in this case north east Peru. The specific objectives were:-

Technically: to demonstrate the flexibility and efficiency of using satellite technology, earth stations, and UHF/VHF radio relay links for rural communications.

Economically: to show that the technical and social installation costs of these technologies to provide rural services are attractively low, and to

TABLE 9.5 Peru's Earth Station Development

Stage	Year	Earth Station	Supplier	Meters	Channels
1	1978	Lurin 2	NEC	13	100
		Iquitoa	NEC	13	56
		Tarapota	NEC	13	28
		Quillabamba	NEC	13	28
2	1981	Chachapoyas	Philips	12	8
		Catallacocha	Philips	12	8
		Contamana	Philips	12	8
		Chavez Valdavia	COMSAT General	8.5	4
		Andoas	Philips	10	4
3	1983	Juanjui	Harris	6.1	4
		Saposoa	Harris	6.1	4
		Tocacho Nuevo	Harris	6.1	4
4	1988	Nationwide TVROs			

demonstrate the feasibility of applying them in other regions with similar characteristics.

Socially: to demonstrate the use of the system in providing social services for education, health and the development of farming skills in the area. At the same time, to assess the socioeconomic impact of the availability of telephone and audioconference facilities, as well as identifying the overall and specific change processes taking place in the towns involved.

(From a paper presented at an ITU Conference held in San Jose, Costa Rica, by Dr Angel Velasquez of ENTEL, Peru in October 1983.)

Users of the telecommunications services provided by the project were initially the health, education, and agriculture sectors, but later on all sections of the community became involved. Through the gradual implementation of the earth stations, ENTEL became familiar with and experienced in the new field of satellite communications, and now have firm plans to expand the earth station network very extensively. The total space segment charge for the preemptible lease from INTELSAT from December 1978 until December 1986 was under US$5.5 million, when the country was leasing 54 MHz at the rate of US$1.02 million per year. In March 1987, Peru purchased transponders 12 and 15 on INTELSAT V, F-3 (108 MHz) for a total of US$1.611 million; the design life of these transponders extends until November 1989. Had Peru continued its 54 MHz preemptible lease, the gross payment for the same period would have been US$2.72 million.

Throughout the development of its domestic satellite system, Peru has taken a cautious and economic attitude in its planning processes. It has made maximum use of INTELSAT's Assistance and Development Program to help in transmission planning and preparation of earth station specifications. It also hosted a seminar held by INTELSAT in Lima for Peru and neighboring countries in 1982 on domestic satellite services.

PHILIPPINES

The Philippines first expressed interest in using satellites for domestic telecommunications services between its many islands in 1975, at about the same time that Indonesia purchased its own PALAPA satellite system. INTELSAT was approached first but the Philippine government decided to lease a transponder on the PALAPA satellite at a cost marginally lower than that offered by INTELSAT. Once equipped for the PALAPA satellites with their different polarization system compared to that of INTELSAT, and the fact that PALAPA is always likely to have excess capacity, only major political problems between the two countries are likely to change this arrangement.

The Philippines have always had an unregulated telecommunications policy that permits private companies to supply telecommunications services. These operate within the framework of the National Telecommunications Commission as part of the Ministry of Communications. These companies include:

Capital Wireless
Eastern Telecommunications (Philippines)
Globe MacKay Cable & Radio
Philippines Communications Satellite Corp. (PHILCOMSAT)
Philippines Long Distance Telephone Company
Philippines Telegraph & Telephone Company

PORTUGAL

Portugal, with its Atlantic Ocean islands of Madeira and the Azores, has been using satellite capacity leased from INTELSAT since 1982 and from the outset stated its intention of increasing its capacity from one and a half transponders in 1982 to two transponders in 1985.

The Portuguese Compania Portuguesa Radio Marconi (CPRM), which is a private company responsible for intercontinental telecommunications services, is also the country's signatory to INTELSAT. CPRM leased 54 MHz of global beam capacity for five years starting in November 1982. Global beams were preferred because the islands are not covered by all the Atlantic satellites and therefore they provide more versatility for the preemptible leased service provided by INTELSAT if it were necessary to transfer services to a different satellite. This lease expired in 1987, when Portugal purchased a 72 MHz east hemispheric beam transponder on the INTELSAT V, F-3 satellite located at 307°E for US\$1.2773 million. That satellite was later relocated in October 1987 and the transponder sold back to INTELSAT. To replace this transponder Portugal purchased the same transponder, number 21, on the next satellite to be located at 307°E which is INTELSAT

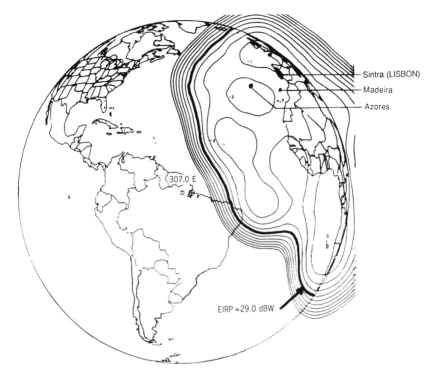

Figure 9.20. Portugal's domestic system beam coverage from INTELSAT at 307° east.

V, F-13. The east hemispheric beam transponder cost US$4.8 million for the entire lifespan of the satellite which is estimated to end in August 1995. The noncancellable leasing cost for the same transponder and period would be US$11.57 million at present day prices. Both satellite transponders provide similar coverage as shown in Figure 9.20.

SAUDI ARABIA

Saudi Arabia's domestic satellite network first started in 1977 using one quarter of a transponder leased from INTELSAT. Various branches of the Saudi Arabian government later leased space segment capacity to suit their individual requirements through the Saudi Arabian Ministry of Posts & Telecommunications, which is the country's signatory to INTELSAT.

Three leases were involved between 1977 and 1986 totalling 81 MHz, two 36 MHz transponders plus the original 9 MHz. The 9 MHz quarter transponder was the first lease (and also the last to leave INTELSAT) and provided a transportable communications network whenever the king traveled.

The remaining 72 MHz leased capacity, which started in 1978, has now been discontinued and transferred to ARABSAT. This provided networks for government and public telecommunications with earth stations at

Riyadh, Jeddah, Jaizan, Sakakah, Sharorah, and Yanbu as well as two transportable earth stations. A new Space Communications City, King Faud City, was opened in 1987 with four earth stations and is the largest single communications complex in the Middle East.

SPAIN

Satellites have been used by Spain since 1975 to relay television between the Iberian mainland and the Canary Islands which are territorially part of Spain. These islands have long been linked to the mainland by cables, but these never had sufficient capacity to relay television programs. Very shortly after the INTELSAT Definitive Agreements came into effect in 1975, Spain asked to use INTELSAT for domestic purposes under the Article III b(i) provision as a country with areas separated by high seas. The original lease was nonpreemptible and the full rate was paid for half a global transponder. This has been maintained since January 1976, although a further half transponder (18 MHz) lease was taken in 1983 to relay a second television channel.

Spain also uses EUTELSAT television services and is also planning its own two-satellite system to commemorate the 500th anniversary of the discovery of America in 1992.

SOUTH AFRICA

Although South Africa has a highly developed national telephone network, television was not introduced into the country at all until 1977, and then only in the main cities. Satellites offered the only viable means of taking the programs to all parts of the country. The telephone network is operated by the South African Post and Telecommunications Corporation (SAP&TC) whereas the television network is owned and operated by the South African Broadcasting Corporation (SABC). SAP&TC is the country's signatory to INTELSAT and encouraged SABC to engineer its domestic television network directly with INTELSAT coordination. The first satellite television channel came into service in 1984 using one 72 MHz hemispheric beam transponder and a further 36 MHz lease is planned for another channel starting in 1989.

SUDAN

Sudan first approached INTELSAT in 1976 asking to lease a full 36 MHz global beam transponder which started operations in January 1977. It is a vast, mainly desert country with the River Nile cutting a green band from north to south along the eastern side of the country. Land transportation was, and remains, difficult, so satellite communications were rightly seen as an opportunity to develop the economy of the country.

The Sudan Telecommunications Corporation (STC) is responsible for telecommunications and comes under the control of the Ministry of Communications. The earth station equipment was procured directly from the U.S. Harris Corporation as a turnkey project. The SUDOSAT system, as it is called, consists of a central control earth station just outside Khartoum and colocated with the international Standard A earth station of Umm Haraz. There are 13 other earth stations around the country as shown in Figure 9.21.

The preemptible lease has been operative since February 1977 and has provided one color television channel and two radio broadcast channels for redistribution purposes as well as up to 100 telephone circuits. All earth

Figure 9.21. Sudan's SUDOSAT domestic satellite system.

stations were equipped for telephone, telex, and the reception of television. The contract also included equipment to rebroadcast the television for local purposes from each earth station. Only the earth stations at Khartoum and Juba were equipped to transmit television through the satellite.

The earth stations were factory assembled in shelters and contained state of the art equipment needing minimum installation when the antenna erection was completed. The SUDOSAT network should have created enormous opportunities for the country, but does not seem to have done so. One of the major problems that seems to have been encountered was the earth station power supplies; they required the U.S. 120 V, 60 Hz supply whereas the Sudanese electricity was nominally 240 V, 50 Hz. This difference meant that every earth station had to be supplied by a diesel generator (like Nigeria). This also proved to be unreliable on account of irregular fuel supplies due to long distances, scarcity, and the high cost of fuel. Consequently, earth station interruptions were commonplace. There were, and still are, constant problems in obtaining spare parts when needed and the overall system has not proved to be successful. After ten years, the earth stations are now nearing the end of their useful life, and it is very doubtful if the system has contributed much to the ecomomic development of the country due to the inadequacies of the earth stations and the way in which they were maintained.

These problems are typical of many lesser developed countries; long distances, rough terrain, lack of reliable electricity supplies, problems in obtaining spare parts, inadequate maintenance training, and hard currency problems.

SWEDEN

Although Sweden has not yet leased or purchased transponders for domestic use from other sources, several Swedish companies have been involved in constructing an experimental multipurpose satellite since the early 1980s, known as Tele-X, which is due for launching by Ariane during the course of 1988. The satellite is designed to be experimental, and to supply all digital services to Sweden, Finland, Norway, and Denmark that will include direct television and radio broadcasting and a variety of data communications services including video.

The satellite, like Germany's TV-SAT and France's TDF, was manufactured by the European consortium, Eurosatellite.

THAILAND

Thailand uses both INTELSAT and Indonesia's Palapa satellites. The main telecommunications organization in Thailand is the Communications Authority of Thailand (CAT) which is responsible to the country's Ministry of Communications and has used INTELSAT for its external services since 1968 and for some internal telephone services since 1982. Another government agency is responsible for broadcasting and has been using Indonesia's Palapa since the late 1970s for television distribution. Other government agencies are showing interest in satellites as a means of developing specialised telecommunications services for state security, public relations, government information, civil aviation, education, and entertainment purposes.

In recent years Thailand has been looking into how best to satisfy all these requirements in the years ahead and has also studied the possibility and usefulness of owning its own satellite. There are certain disadvantages to leasing facilites in different satellite systems as happens at the present time, especially when for technical differences, the earth stations used cannot be used with both satellites. The range of requirements are wide and to meet them it may well be necessary to take the most costly approach and for Thailand to purchase its own satellite. In November 1986 there were reports that Thailand was in fact discussing with the U.S.S.R. the possibility of launching an RCA satellite. There were obvious political problems with this approach as the United States is generally not in favor of permitting its technical know-how to be available for inspection and use by such countries as the U.S.S.R. and there are legal problems in such taking place. However, it seemed that Thailand had taken at least a preliminary decision to proceed independently towards its own satellite system. This proved correct early in 1988 when Thailand called for bids to launch and operate a satellite over the next 30 years that would permit Thailand to use any excess capacity for its own purposes, including resale of transponders on a regional basis.

As a precurser to expanding satellite capacity two Thai organizations (Sophonpanich and Srifuengfung) have teamed up with the Cable & Wireless Group in a joint venture to provide state of the art satellite data communications services.

TONGA

This remote Pacific island kingdom would seem another unlikely candidate for its own satellite system but late in 1988 it applied to the IFRB for eight orbital locations for its future satellites, and its intention to operate three "Tongasat" satellites starting in the early 1990s. Tongasat is the name given to the satellites of the Freindly Islands Satellite Co. The action of applying to the IFRB effectively prevents other applicants from using these locations, unless satellites are not launched before 1994. A later statement revealed that the remaining orbital locations could be used by other satellite operators on a leased or purchased basis. No firm plans have been announced regarding the Tongasat satellites but much of their capacity will need to be used by countries other than Tonga.

TURKEY

Turkey first started considering the use of satellites for television distribution in 1984 and was able to consider leasing from either INTELSAT or EUTELSAT, being members of both organizations. One of the prime motivations for this need was that Turkey had been receiving the U.S. AFRTS program from INTELSAT's international television leased transponder, and there were many internal requirements for a wider distribution of the program. There was also the political need to distribute Turkish national television as it would obviously be unacceptable to distribute foreign programs without distributing national television in the same manner.

The Turkish PTT decided to purchase two Ku-band transponders on the west spot beam of INTELSAT V, F-7, located at 66°E for a total cost of US$5.989 million. Transponder 69 was purchased for US$3.251 million starting in August 1986, and transponder 69 was purchased for US$2.738 million starting in April 1987, both transponders being expected to last until at least September 1990. Had the same transponders been a noncancellable lease for the same period, the cost to Turkey would have been nearly US$14 million gross, or approximately US$8 million net when the INTELSAT financial arrangements are taken into account.

UGANDA

Uganda was the least successful of any country adventuring into the use of satellites for developing a country's ecomomy. Uganda, under the Presidency of Idi Amin, leased a full 36 MHz global beam transponder from INTELSAT starting in August 1977 until April 1979. As a system it was ill-conceived and from the start offered almost nothing of value to the country of Uganda; in 1979 the Uganda authorities decided to ask INTELSAT to allow termination without penalty. Much to its credit, INTELSAT noted that the situation resulted from deplorable events that were beyond the signatory's control. Uganda did not use the space segment during the period of its full-time lease and made only very minimal use of the space segment during the whole period. INTELSAT considered that the country had paid much more for the service than they had actually used and therefore INTELSAT exonerated Uganda and cancelled all outstanding lease charges.

A map of Uganda is shown in Figure 9.22 together with a network of six domestic earth stations that had been proposed to the Uganda government. The system proposed the village of Arua, some 300 km from the capital city, Kampala, as the domestic system center, and also the international gateway. Arua's only claim to fame at that time was that it was the home of President Idi Amin. Fortunately, a Standard A international gateway station was

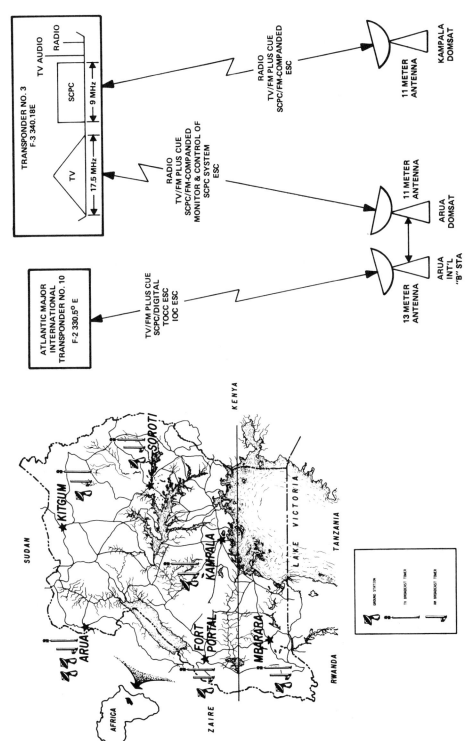

Figure 9.22. Uganda's domestic satellite network in 1978.

already being established in Kampala so that when the domestic system finally died a natural death, international communications could still take place. Only three earth stations were actually installed and all were totally written off within a year or so.

UNITED KINGDOM

The United Kingdom has been considering its own domestic satellite system for many years but it was not until December 1986 that the country's Independent Broadcasting Authority (IBA) selected an organization to provide the satellite services. The United Kingdom had once been the most likely runner to be the first European country to introduce a DBS television service, but political differences kept creating delays. In the early 1980s there was great enthusiasm in creating widespread cable and direct satellite television services, but system costs and legislative haggling slowly dampened the enthusiasm. There were many diverse interests involved, and at the start the government gave responsibility for the satellite development to the British Broadcasting Corporation (BBC). A consortium of British Aerospace, British Telecom, and the General Electric Company (GEC), named United Satellites Ltd. (Unisat), was created to operate the satellite system on behalf of the BBC. Unisat contracted for three satellites manufactured by British Aerospace and GEC/Marconi that would contain facilities for both DBS television and fixed satellite services, with some of the transponders being offered to INTELSAT. Construction of the satellites was stopped in 1983 when the BBC decided that it would not be able to operate the DBS service commercially. The government then asked the IBA to join with the BBC to develop a DBS plan jointly, and another consortium of 21 companies, named BritSat, was formed to study and produce a satellite system. A number of satellite proposals were studied but, once again, the outcome of the study was that a DBS service would not be commercially viable as lease charges per transponder would be in the order of US$120 million. Additionally, the preferred option for BritSat's RCA satellites caused problems with British industry which had already invested a lot of money into the partly built UniSat satellites. Further problems arose in 1985 when the BBC, intended as a 50% user, claimed that it could not provide a DBS service without further funding, and also the government of Ireland was giving consideration to an Irish DBS TV system that would also cover the whole of the British Isles.

In December 1986, both the British and Irish governments approved their different plans. The British IBA selected a new consortium, British Satellite Broadcasting (BSB), and provided it with a 15 year license to operate a DBS system. BSB is a consortium of Granada Television, Anglia Television, the Virgin Group, Pearson (a diversified media company), and Amstrad (a British consumer electronics group). BSB expects to start transmitting three

DBS channels for the United Kingdom in 1990 at an estimated total cost of US$850 million. Additional investors are also being sought and foreign investment is not excluded to raise the estimated US$725 million needed to get the project into orbit.

BSB chose General Motors' Hughes Aircraft to supply the satellites that will have 16 TV transponders similar to those used by Luxembourg's ASTRA. Many more problems could occur before the United Kingdom gets its own satellite and DBS television. The British are generally avid television viewers, and BSB expect 400,000 private dishes to be sold during the first year of service and to about half of the United Kingdom's 20 million homes by the turn of the century. Private home-receiving sets are expected to be about 40 cm in diameter and cost under US$500 each from British manufacturers. Square (25 cm) rooftop antennas are also being marketed.

British television has historically centered on quality rather than quantity and the final selection process seems to indicate little change; there will not be a multiplicity of channels as exists in the United States where many channels are repetitive and duplicate each other.

While all the planning was taking place, KU-band INTELSAT capacity has been used on a leased basis since 1984, and further Ku-band capacity will be leased on Luxembourg's ASTRA satellite until BSB's own satellite is launched in 1989. TV channels are presently transmitted from London's Teleport and received by about 50 TVROs for local rebroadcast or cable TV services.

PanAmSat also found its way into the UK domestic satellite communications field in 1988 in the wake of an increasing demand for further deregulation of the country's telecommunications; the government permits no more than six licences for point-to-point services.

THE UNITED STATES

The first U.S. company to seek FCC authorization to build a domestic satellite system was the American Broadcasting Corporation Inc. (ABC) in 1965 for distributing television. This prompted a debate of policies by numerous U.S. government agencies that continued for several years and may even be said to be still going on. The Nixon administration, in 1970, through its Office of Telecommunications Policy recommended an "open entry" policy and this policy has not changed. Many companies have aspired to become satellite owners and operators, of which many did not fulfill their aspirations. It was foreseen in 1972 by the OTP that at least two, and probably three or more, separate systems would be established, having a combined capacity in excess of 100 transponders, plus 50 or more backup transponders. In 1986 there were eight companies owning and operating satellite systems, although numerous others are planning to do likewise. The satellite operating companies late in 1988 were:

Company	Since	In Orbit
Western Union Telegraph Co. (WUTCO)		
(now GM's Hughes communications)	1974	3
GE's RCA Americom	1975	4
Alascom	1982	1
COMSAT General (COMGEN)	1976	3
MCI's Satellite Business Systems (SBS)	1980	4
GM's Hughes Communications	1983	3
AT&T	1983	3
General Telephone & Electronics (GTE)	1984	6
CONTEL's AMSAT	1985	2
	Total	29

Of the nine domestic systems operating in late 1987, according to the FCC, at C-band there were 414 transponders with an overall fill factor of 67%, and at Ku-band there were 122 transponders with a 57% fill factor as detailed in Table 9.6.

The actual use of communications satellites by the United Sates at the end of 1986 was much less than had been authorized by the FCC three years earlier in 1983, when many companies foresaw large profits to be made in communication satellite applications. The 1983 FCC authorizations are shown against the 1987 acutals in Table 9.7 and although some of the shortfalls can be blamed on launch problems following the shuttle disaster in January 1986, much is also due to the demand forecasts being over optimistic and other economic factors. An FCC report early in 1988 stated that satellite over capacity increased during 1987.

Only a general synopsis of United States satellite communications applications is given because the situation is continually changing as new companies enter the field, companies leave the field, and mergers, takeovers and name changes occur. The eight existing systems are described at some length but other possible aspirants are only briefly mentioned. Although beam coverages of the satellites differ somewhat, they tend to have many similarities:

Continental U.S. coverage (CONUS)

Coverage of Puerto Rico and Caribbean area

Spot beams for Hawaii and Alaska

These are shown in Figure 9.23.

Having one of the oldest and most respected names for telecommunications in the United States, it was not surprising that the Western Union Telegraph Company (WUTCO) should have been the first company to own and operate its own satellite network by launching its first two satellites, WESTAR 1 and 2, in 1974, and then launching its ground spare in 1979

TABLE 9.6   United States Satellite Systems in 1986

Company/Satellite	Year of Launch	Band	No.	MHz	°W	Inactive	% Fill
			Transponder Location			Efficiency at 1987	
AT&T Telstar 301	1983	C	24	36	97	6	65
302	1984	C	24	36	85	5	66
303	1985	C	24	36	126	7	69
CONTEL AMSAT 1	1985	C	18	36	128	1	73
		Ku	6	72	128	1	83
2	1989	C	18	36	81		
		Ku	6	72	81		
COMGEN Comstar 1	1976	C	24	34	Retired 1984		
2	1976	C	24	34	Retired 1984		
3	1978	C	24	34	Retired 1984		
4	1981	C	24	34			
GTE Gstar 1	1986	Ku	22	54	103	6	77
2	1985	Ku	22	54	105	7	50
3	1987	Ku	22	54	Launch failure		
3R	1989	Ku	22	54	?		
Spacenet 1	1984	C	24	36	120	7	35
		Ku	6	72	120	2	32
2	1984	C	24	36	69	7	75
		Ku	6	72	69	7	50
3	1987	C	24	36	Launch failure		
		Ku	6	72	Launch failure		
Hughes Galaxy 1	1983	C	24	36	134	1	100
2	1983	C	24	36	74	9	38
3	1984	C	24	36	93	0	100
4	1989	C	24	36	140		
RCA SATCOM 1	1975	C	24	36	Retired 1984		
2	1976	C	24	36	Retired 1984		
3	1979	C	24	36	Launch failure		
3R	1981	C	24	36	130	1	95
4	1982	C	24	36	83	2	89
ALASCOM 5	1982	C	24	36	142	13	45
6(1R)	1983	C	24	36	138	13	61
7(2R)	1983	C	24	36	72	1	65
K1	1985	Ku	20	81		12	25
K2	1985	Ku	20	81		4	60
SBS/MCI SBS 1	1980	Ku	10	99		0	90
2	1981	Ku	10	97		1	70
3	1982	Ku	10	95		4	55
4	1984	Ku	10	91		6	40
5							
6							
WUTCO Westar 1	1974	C	12	36	Retired 1983		
2	1974	C	12	36	Retired 1983		
3	1979	C	12	36	91	2	72
4	1982	C	24	36	99	3	68
5	1982	C	24	36	124	6	55
6	1984	C	24	36	Lost and recovered		
6R							

TABLE 9.7   U.S. Satellite Authorizations and Actuals[a]

Company	1983 Authorized		1983 Pending		1987 Actuals	1987 Applications
	Build	Launch	Build	Launch		
Western Union	6C	6C	11C 6Ku	11C 6Ku	3C –	
GE/RCA Americom	9C 4Ku	8C 3Ku	4H	3H	4C 2Ku	2H 1Ku
GE/RCA Alascom	1C	1C	3C	2C	1C	1C
Comsat General	4C	4C	3Ku	3Ku	1C	
GM/Hughes Galaxy					3C	1C, 2Ku
SBS	5Ku	5Ku	4Ku	4Ku	4Ku	
AT&T	3C	3C			4C	3H
GTE	3Ku	2Ku	1Ku	1Ku	2Ku	1Ku
GTE Spacenet	4H	3H		1H	2H	3H
CONTEL/AMSAT	3H	2H	2H	2H	1H	3H
Advanced Business	3Ku	3Ku				
Rainbow	3Ku	2Ku	2Ku	2Ku		
U.S.S.S.	3Ku	2Ku	2Ku	2Ku		
Cablecast General			3C	2C		
Columbia			2H	2H		
Digital Telesat			1C 3Ku	1C 3Ku		
Equatorial			3C	3C		
Federal Express			3Ku	2Ku		
Ford			4H	3H		
Martin Marietta			3Ku	2Ku		
National Exchange			3C	2C		3H
Total	51	44	63	57	27	20

[a] C, C-band satellites: Ku, Ku-band satellites; H, hybrid satellites with both C- and Ku-bands.
Source: Aviation Week 5 October 1987.

when the system had grown. Despite its 14 year background in the satellite business, WUTCO, financially troubled, sold its satellites to General Motors' Hughes Communications early in 1988.

WESTAR 1, 2, and 3 were supplied by the Hughes Aircraft Company and apart from antenna beam patterns were identical to Canada's ANIK and Indonesia's Palapa satellites. They were all designed round the INTELSAT IV series having twelve 36 MHz transponders, but used linear polarization as opposed to INTELSAT's circular polarization. WESTAR 1 and 2 were retired in 1983 after serving nine years instead of the expected seven years and they were replaced by WESTAR 4 and 5 which are compatible C-band satellites but which have 24 dual polarized transponders thus doubling the capacity of the earlier satellites.

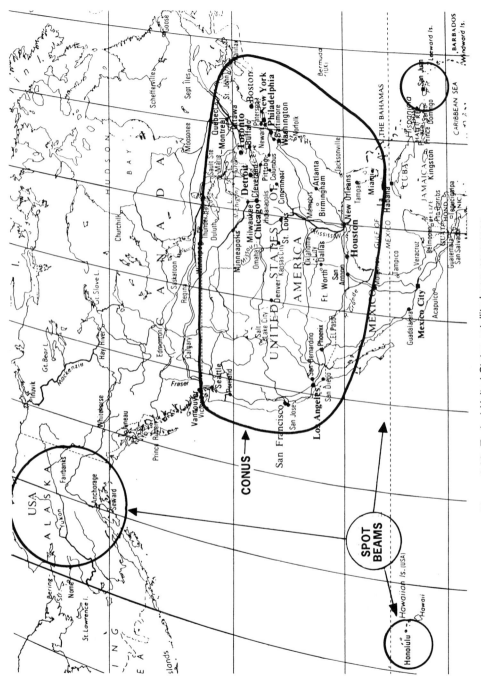

Figure 9.23. Typical United States satellite beam coverages.

253

WUTCO now has three satellites in use in orbit:

WESTAR 3 at 91°W
WESTAR 4 at 99°W
WESTAR 5 at 123°W

A fourth satellite, WESTAR 6, was launched from the NASA shuttle on 3 February 1984 with Indonesia's Palapa B2 satellite, but both satellites failed to reach geosynchronous orbit due to failure of the Payload Assist Modules (PAM). Both satellites were successfully retrieved from space by a later shuttle mission. WESTAR 6 had cost US$75 million but, to cover loss of business was insured for US$105 million. This was later sold and acquired by Asiasat.

The satellite system is controlled from the company's main earth station at Vernon Valley in New Jersey, and there are other major earth stations near Atlanta, Chicago, Dallas, and Los Angeles. WESTAR antenna patterns cover the continental United States mainland with a Conus beam and use spot beams to Hawaii. The Conus beams also provide about 30 dBW to most of Mexico and the Caribbean Islands and part of Canada. Other U.S. domestic satellites are similar, but it was WUTCO who first raised the transborder issue of providing satellite services internationally to neighboring countries.

During the late 1970s, the Spanish International Network (SIN) wanted to lease a WESTAR transponder for television services between Mexico and the United States which would have been an infringement of the INTELSAT Agreements by both Mexico and the United States. To overcome this, an arrangement was reached whereby INTELSAT leased the transponder from Western Union and in turn leased it to the signatories of Mexico and the United States at no added cost.

Many of the other services on WESTAR satellites are data networks and television distribution services, all of which could be extended to the Caribbean islands were it legal to do so. Consequently, many receive-only earth stations can be seen at many Caribbean and Central American locations for receiving U.S. television programs. Many data networks, however, need the ability to transmit, and in the case of the Caribbean, this means international service which again infringes the INTELSAT Agreements. Although many of the smaller caribbean islands are not INTELSAT members, the expansion of data networks into the Caribbean has not yet transpired by means of the United States private domestic satellites. The whole question of transborder services from the United States to neighboring countries remains a maze of political and legal issues, and Western Union, with others, can be expected to make any inroad wherever possible.

A typical television distribution application is that of the Public Broadcast System (PBS) use of WESTAR 4 transponders. PBS is a private, nonprofit

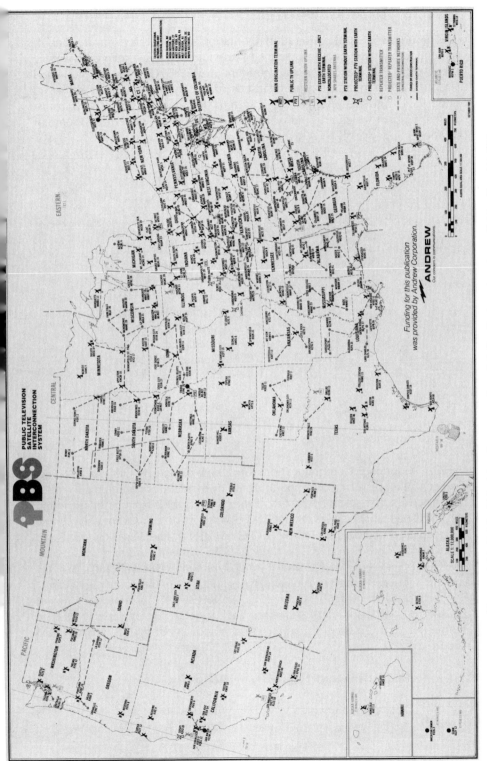

Figure 9.24. Public Television Interconnection Satellite System (courtesy: Public Broadcasting Service).

corporation owned and operated by the nation's public television stations. It was founded in 1969 and now claims to provide the largest television service in the world. PBS leases four transponders on WESTAR 4 for collecting and distributing its programs to its hundreds of television stations as shown in Figure 9.24. The present WESTAR system has seventy-two 36 MHz transponders most of which are sold or leased to other service operators; some of the applications of WESTAR and the users include:

Radio Broadcast Networks	PBS
	CBS
	ABC
Television Networks	PBS
	Wold Communications
	Catholic Television
	Bonneville
	ABC
	CBS
	Hispanic Broadcasting
	National Christian Net
	Unchannel TV
	Japanese TV News
	ProAm Sports
	Group W TV
	Meadow Racing
	Fantasy TV
Business Networks	Dow Jones

RCA Americom Communications Inc., now owned by General Electric, has been manufacturing and operating its own satellites commercially since 1975 when its first series SATCOM 1 and SATCOM 2 became operational. The third one of the series, SATCOM 3 was lost on launch in 1979. These were replaced and augmented by SATCOM 3R, 4, and 5 in 1981/1982 and SATCOM 6 and 7 in 1983; a total of four C-band satellites in service. SATCOM 5 was sold to Alascom Inc. in 1982 for US$88 million to provide satellite services within Alaska and to connect it to the remaining States. RCA operates the satellite, known as AURORA 1, on Alascom's behalf and four of its transponders are leased back to RCA for six years at US$1.6 million per year. Some of the applications and users of the RCA SATCOM system are:

Radio Broadcast Networks	ABC
	NBC
Television Networks	Nickolodeon
	Bravo
	ABC

Home Sports
Sports Vision
New England Sports
Playboy
Silent Network
TBS Superstation
Odessey
Cinemax
Arts & Entertainment
NETCOM
AFRTS
RAI, Italian
Business Networks   Aurora
NASA Control

During the peak days of COMSAT, its subsidiary COMSAT General (COMGEN) launched satellites for leasing to the two major U.S. carriers AT&T and GTE. These satellites COMSTAR 1, 2, and 3 launched between 1976 and 1978 were retired in 1984 and were not replaced. However, in 1986, COMSAT announced its ability to resurrect them using an inclined orbit maneuvering technique that could prove useful if launching problems persist.

The Hughes Aircraft Company has been, and remains, one of the world's leading communications satellite manufacturers since the early 1960s. Hughes Communications was formed in 1978 as a wholly owned subsidiary of the Hughes Aircraft Company, and both are now fully owned by the General Motors Corporation (GM).

Hughes also has financial interests in the Japanese Communications Satellite Company and Ireland's Atlantic Satellites which together could provide formidable competition to other international service suppliers in the United States as well as INTELSAT should the three combine Europe, the United States, and Japan. As already mentioned, Hughes communications has also purchased the WESTAR satellites.

The company has been providing satellite services through its three C-band Galaxy satellites and associated earth stations since 1983. Prior to that it has been involved in government and satellite support projects and services on a leased basis using its SYNCOM and LEASAT satellites.

At the present time, all the Galaxy satellites in orbit are C-band although Ku-band satellites are also planned for the future. Galaxy 1, at 134°W is used entirely for television distribution leased services. Galaxy 2 and 3 located at 74° and 95°W, provide business and video services, but are only about half filled.

Hughes' next generation of satellites, likely to be launched in the early 1990s are planned to provide 16 Ka-band transponders to 16 major cities using individual spot beams to each city.

The American Telephone and Telegraph Company (AT&T) had been involved with satellite communications even prior to the formation of COMSAT. Its first experimental satellite was called TELSTAR 1 and provided the first commercial telephone and television services over the Atlantic from Andover, Maine to earth stations in France, United Kingdom, and West Germany. AT&T was the first major user of INTELSAT services through the United States signatory to INTELSAT, COMSAT.

In 1976, COMSAT General (COMGEN) launched two domestic satellites COMSTAR 1 and 2 followed by COMSTAR 3 in 1978 and COMSTAR 4 in 1981 which were mainly leased to AT&T for U.S. domestic long distance telephone purposes. AT&T called these the TELSTAR 2 series. The company's new domestic satellites are also called TELSTAR and are designated 301, 302, and 303; they were launched starting in 1983. TELSTAR 4 is a ground spare. These now provide AT&T services to all 50 states and Puerto Rico although some transponders are leased out for television distribution services.

TELSTAR 301, 302, 303, and 304 are all Hughes C-band satellites equipped with twenty-four 34 MHz transponders using dual linear polarization and providing a down-link EIRP of about 33 dBW. AT&T's main satellite control station is located at Hawley, Pennsylvania with alternate control possible from the Three Peaks facility in California. The satellites are expected to provide useful service well into the 1990s. Apart from its own C-band satellites, AT&T also leases two Ku-band transponders on RCA's SATCOM K2 satellite for providing its Skynet business data services. For the future, AT&T agreed to take control of Ford Aerospace Satellite Services for about US$27 million which includes licenses to own and operate two Ku-band satellites around 1992.

Satellite Business Systems (SBS), now owned by MCI, was formed in 1977 by a partnership of COMSAT General, International Business machines (IBM), and the Aetna Life and Casualty Insurance Company to provide satellite communications facilities on a common carrier basis in the United States. The new company chose the Hughes Aircraft Company to build three Ku-band satellites known as SBS-1, 2, and 3, the first of which was launched late in 1980 to become the United States' first commercial Ku-band satellite. Services were targeted towards private business networks, government agencies, and other organizations with large communications requirements. SBS used all digital, TDMA transmission techniques.

The first five years of operation were lean with heavy financial losses of over US$400 million, and there were several changes in ownership structure with its future still remaining uncertain under the ownership of MCI. There are four satellites in orbit; three owned and operated by MCI, and the fourth owned and operated by IBM through a marketing subsidiary. Two additional satellites were to be launched in 1987 and 1989, but system construction by Hughes has been delayed due to launch and business problems. An attempt was made to sell off the partly constructed SBS-5 satellite to INTELSAT late in 1986 which indicates SBS reluctance to create more in-orbit spare capacity at an expense which its new owners can ill afford in the highly competitive market that now exists in the United States.

Another unusual aspect of SBS was that it developed its own earth station designed for office premises with TDMA transmission techniques to provide integrated voice, data, facsimile, and teleconferencing services to customers all over the country. After an international selection process, Hughes and Japan's NEC each supplied 100 earth station terminals to start off the system. As its name suggests, SBS satellites are mainly applied to business networks, now mainly operated by MCI, but SBS-3 is predominantly used for television networks indicating the fact that the business demand was not up to original expectations.

General Telephone & Electronics (GTE) has two domestic systems, the GSTAR system and the Spacenet system. The GSTAR system was devised by the GTE Satellite Corporation whereas the Spacenet system was originally procured from RCA Astronautics by the Southern Pacific Communications Corporation which was later purchased by the GTE group. The GSTAR satellites are Ku-band only, whereas the Spacenet satellites are hybrid using both C- and Ku-bands. Spacenet 3 was destroyed during the launch failure of its Ariane rocket in September 1985, and GTE Spacenet was authorized, at least on a temporary basis, to use the GSTAR facilities. GTE's principal service is called Skystar which provides multipoint one-way data, point-to-point interactive, or international business data services. Like other carriers, GTE's Skystar international service passes to INTELSAT via a GTE owned earth station.

Private business networks are also provided and the first large user was K-Mart which involved a cost of about US$40 million including the purchase of over 2000 very small earth stations as well as a hub and network control facilities. This network is now being installed and will shortly allow the K-Mart headquarters to communicate with its 2000 nationwide stores more quickly, efficiently, and economically than was hitherto possible. This project is being performed by U.S. Sprint's Telenet, using very small earth stations supplied by NEC America, the same company that supplied earth staions for the SBS business networks.

The American Satellite Corporation (AMSAT) was first formed in 1974 and is now a wholly owned subsidiary of the Continental Telephone Company (CONTEL) since CONTEL bought out AMSAT's earlier partner, Fairchild Industries, in 1985. The original corporation was the first new company, in 1974, to be formed with the objective of providing competition to the existing COMSAT monopoly in satellite communications.

While AMSAT was organizing itself and its satellites, which took until 1985 before the first was launched, transponders were purchased on the WESTAR system. Like other U.S. domestic satellite systems it includes coverage of Alaska, the Hawaiian Islands, and Puerto Rico as well as the continental United States.

CONTEL/AMSAT targets its services towards the large business corporate users which presently include Xerox (see also Chapter 5), Federal Express, the New York Times, Metropolitan Life Insurance, Bank of America, and United Technologies. As well as business corporations, CONTEL also has a US$16 million contract with NASA for one full transponder for services linking 14 NASA earth stations with the Vandenburg Airforce Base in California. A recent corporate feature of CONTEL/AMSAT is that it is also trying to offer services outside the United States. In 1986–1987 it tried to start a direct satellite service between the United States and Montego Bay, Jamaica. This is technically possible because, like other domestic satellites, the beam coverage also includes most of the Caribbean and Central American countries. The proposed service was to use the cheap labor available in Jamaica to substitute for more expensive labor in the United States for employment at a large telemarketing/home shopping hub for large United States toll-free-number services. The proposal was attractive to the Jamaican government because it would have provided some much needed extra employment, and to the United States customers it would have also provided a source of much less expensive labor. It was also a wedge in the door for entering the international arena using spare U.S. domestic capacity.

It was claimed that the service could be provided much more economically than by INTELSAT and that were INTELSAT to be used, the Jamaican end of the service would not be viable. It was further claimed that "end user charges for services provided via the INTELSAT system for international voice circuits will never fall to the level where they would eliminate the sharp price differential between domestic U.S. services and current end-to-end INTELSAT charges. A single voice-grade circuit between Kingston, Jamaica and the mid-western States costs the end-user $4,300 per month. Even if such charges are reduced substantially in the future, it is unlikely they will reach the $600 to $1,000 per month rate which Teleport is able to charge its customers. If faced with no alternative to INTELSAT, all voice service providers would move back to the United States." The claim was made by AMSAT through the U.S. signatory to INTELSAT, and the government of Jamaica.

The actual INTELSAT cost for an unmultiplexed circuit between Jamaica and the United States of this nature is between US$725 and US$925 per month, and all other charges are payable by the end-user to the Jamaican and U.S. signatories. On the assumption that the end-user cost for one circuit between the Midwest and Jamaica is US$4300 as claimed, then the breakdown cost must be:

INTELSAT	US$ 900
COMSAT	US$1700
Jamaica	US$1700

There was thus a wide discrepancy in costs being quoted within the U.S. policy-making process. As it turned out, CONTEL/AMSAT satellites were not used as at a later FCC hearing, COMSAT claimed that it could provide a multiplexed INTELSAT circuit to provide as many as five suitable voice circuits at an effective price of about US$312 per month compared to CONTEL/AMSAT's US$575.

CONTEL/AMSAT's services include a full range of voice, data, facsimile, and video applications through about 200 earth stations nationwide. Satellite and network control is performed from the company's TTC&M earth station near Atlanta, Georgia:

Flexstream service, integrated data services up to 1.544 Mbps (T-1)
Dedicated private networks
Video leases
Videoconferencing
IBS international services via INTELSAT
National networks
Government services

During the last six or seven years many new companies have seen a glowing future for owning and operating satellite systems. Many of these companies have not lasted the course, and as discussed elsewhere, many more are unlikely to survive, but some (not all) of these companies are listed below to give the reader an idea of the extent of market interest that existed.

Chief Application	Name of Company
Business Networks	Advanced Business Communications
	Ecuatorial Communications
	International Satellite Inc.
	ORION
	National Exchange (SPOTNET)
	Space Communications Services
	Columbia Communications
	Terasat
	PanAmSat Corp.
	Pan Am Satellite Corp.
Direct Broadcasting (DBS)	Advanced Communications Corporation
	DirectSat Corporation
	Direct Broadcast Satellite
	Echostar Satellite Corporation
	Orbital Broadcasting

Chief Application	Name of Company
	Tempo Satellite Inc.
	Continental Satellite Corporation
	Satellite Syndicated Systems
	Hughes Communications
	U.S. Satellite Broadcasting
Mobile Applications	Skylink Corp.
	Mobile Satellite Corp.
	McCaw Space Technologies
	Wismer & Becker Transit

The United States domestic satellite telecommunications market is the world's largest and is likely to remain the most subject to competitive influences. Telecommunication service providers are likely to continue to change their names, merge, and emerge during the years ahead.

U.S.S.R.

It is well remembered that SPUTNIK I was the world's first artificial satellite when it was launched on 4 October 1957. This success immediately stirred the United States into action but the second SPUTNIK was launched before the first U.S Explorer launch took place in January 1958. It was the tenth SPUTNIK launch on 25 March 1961 that put the first man (Yuri Gagarin) in space.

The first satellites launched for communications services were the Molniya series (meaning lightning) which were not geostationary. These were first launched in 1965 to broadcast radio and television services from Moscow to towns and villages throughout the U.S.S.R. Molniya satellites track a highly elliptical orbit and required earth stations to track them as they orbit the earth. To the U.S.S.R., this type of satellite has many advantages over the geostationary satellite because they can be used for the polar areas of the country that are too far north to see geostationary satellites at the equator. The system is still in use today for television distribution services.

The first geostationary satellite (Statsionar) was launched in December 1975, and over 40 were launched in the next ten years. The operational life of the earlier versions was only about two years compared to the Western norm of seven years. The general soviet policy seems to call for a greater number of different satellites for more specific uses whereas Western satellites tend to be larger multipurpose satellites.

The Raduga satellites were the U.S.S.R's first geostationary communications satellite launched in 1975. The Raduga (meaning rainbow) satellites have been used to provide telephone, telegraph, and domestic television services from orbital locations at 35°, 40°, 85°, 128°, and 35°E. Little is really known about the technical parameters of the basic design, but they are believed to be modeled after the Western three-axis stabilized platform.

The Elkran satellites (meaning screen) are used only at 99°E and provide direct broadcast services to transmit television to the far north of the U.S.S.R. They have a 200 W transmitter with a very narrow beam antenna to allow reception by small, simple earth stations in villages and on individual buildings.

Gorizont (meaning horizon) satellites have considerably higher down-link power than most Western satellites, presumably to provide better news and television broadcasting requirements for voice and commercial traffic.

Most Western C-band satellites transmit less than 10 W of power per transponder, whereas Gorizont satellites have 15–40 W TWTAs with global, hemispheric, zone, and spot beam capabilities with each coverage pattern selectable by control from the ground. Power levels range from a minimum of 26 dBW at the coverage edge of the global beam to a maximum of 48 dBW at the center of the steerable spot beam. This higher power satellite could prove very attractive to lesser developed countries that had difficulty in persuading INTELSAT to adopt a similar approach. If the U.S.S.R. markets these services seriously, it could make considerable inroads into the INTELSAT market.

Like many U.S.-made satellites, they use 36 MHz transponders in the C-band frequency range. Their beams are contoured for global and northern hemispheric coverages and use circular polarization with somewhat higher down-link power than the INTELSAT satellites.

Loutch (meaning ray) satellites are the most recent series of Soviet communications satellite packages and each carries one Ku-band transponder used mainly for telephone and data services.

A future series of satellites called TOR will be introduced in 1990 and 14 are planned to provide almost full global coverage. It is reasonable to suppose that these will also greatly increase the global scope of INTER-SPUTNIK. The TOR satellites will use a new 42.5–45.5 GHz up-link and 18.2–21.2 down-link for fixed land services, as well as mobile services for land, aeronautical, and maritime uses. These satellites look like offering a wider range of services than those of INTELSAT and INMARSAT combined, at least for the world's communist affiliates. This may attract more countries to the INTERSPUTNIK system than there are at present.

VENEZUELA

Venezuela has been leasing 36 MHz of satellite capacity from INTELSAT since 1982. The capacity was used to distribute one channel of television from Camatagua, the country's international gateway earth station near Caracas, to San Fernando de Apure and Puerto Ayacucho, towns on the Orinoco River, and also to San Telmo, a resort mountain town.

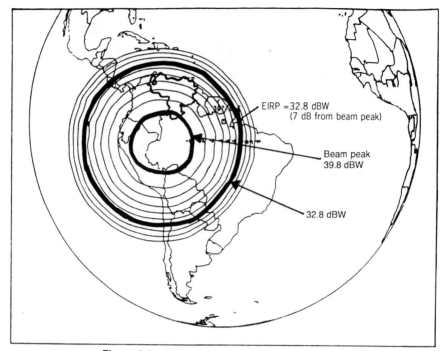

Figure 9.25. Venezuela's domestic satellite system.

Two years after the network started, in 1984, the government saw the need to expand its use of satellites to expand cultural, social, and economic growth within the smaller towns. This meant expanding the existing television distribution with many more TV receiving stations and also, possibly, to add more television channels. A network of small earth stations was also needed to link small towns to the nation's capital using a thin route communications network. Like many other countries, Venezuela used the services of INTELSAT's Assistance and Development Program to develop the best method of achieving the government development objectives.

Venezuela decided to purchase transponder 37 on one of INTELSAT's new C-band steerable spot beam transponders on INTELSAT V, F-13 subject to successful launching late in 1987 or 1988. The cost of the transponder for its estimated seven-year life will be US$4.489 million, and can be expected to provide considerably more satellite power than has hitherto been available. The new beam contour coverages of this steerable spot are shown in Figure 9.25.

ZAIRE

Zaire, another enormous country, predominantly jungle, but with vast natural resources, started leasing a full 36 MHz global beam transponder from INTELSAT in 1978. The 36 MHz has been used to distribute one channel of national television and to provide trunk telephone channels between the capital, Kinshasa, and 13 other towns nationwide. The intention has always been to use these 13 towns as hubs of further local networks and gradually extend the satellite network into rural areas. Although this expansion is still planned, the country's continuing financial problems and difficulties in obtaining funds, or hard currency, continue to delay any positive action.

10

A VIEW TO THE FUTURE

In the early 1980s there was unabashed optimism for a glowing future in all fields of satellite communications, especially in the United States where many new companies emerged to secure a piece of the action. After the pendulum rose, the pendulum dropped again, and many that earlier saw enormous profits now see a somewhat chilly future. Some have realized that satellite communications can still be a risky business, and that rosy pictures of the market place are not always what they first seem.

Some of the glitter may have gone does not diminish the prospects for the future. There is still an enormous untapped worldwide market place for satellite communications services, profitable and unprofitable. This final chapter is just one person's view of where the growth of satellite communications applications may be changing or emerging.

10.1. COMPETITIVE INFLUENCES

10.1.1. Deregulatory Trends

The United States' deregulatory policy, started in the late 1960s, has caused major changes in that country's telecommunications practices and structure. Whatever the merits or demerits of the policy, the same policy is now being pressed successfully in other countries and also into the international arena. The reader will recall that about 100 years ago most of the telecommunications companies were also private, unregulated companies. They pioneered the industry and then needed government regulation to control them, and later became government agencies or administrations. If many of the newly proposed satellite operating companies come to fruition, then it is certain that regulation will again be necessary both domestically and internationally.

Elsewhere, private companies are emerging (or reemerging) in Europe, Japan, and elsewhere to supply services hitherto available only through government-owned companies. These private companies are not likely to proliferate to the same extent that they were allowed to do in the United

States, but they are likely to provide momentum to the slow moving government monopolies in providing a wider variety of applications and services to the end-user at a lower cost.

These trends have so far been beneficial to the telecommunications satellite industry and are encouraging the introduction of new services as well as increasing the volume of traditional services domestically and internationally. The deregulatory trend internationally necessitates private competition to the international organizations that were set up like treaties. In these, large degree of government control is inherent and it is not so easy to introduce as much competition when at least two participants are needed to bypass such a treaty. Despite all the rhetoric in favor of introducing such competition, there are grave economic risks involved in creating a new satellite system, and then being unable to use all its capacity. Such risks must be based on economic reality and the foreseen requirements must culminate as being real requirements with a real revenue base, paid for in real money. Providing satellite services to countries, organizations, or industries that cannot afford the service is not good business in the long-term. Once these systems become closer to reality, second thoughts will emerge and some companies will withdraw themselves without any regulation or further deregulation. The "open entry" telecommunications policy also permits the entrant to exit discretely through the same door, and the future may well see more exits than new entries.

The government attitude in the United States will continue to influence the world satellite communications environment, not only through the large United States' investment shares in INTELSAT and INMARSAT (approximately 25% and 28%, respectively), but also through its high world trade investment. This is a natural and healthy situation so long as the government policies towards these organizations are responsible and consistent; the United States' representation at these organizations should present a U.S. policy that is seen to be cooperative and consistent. The same is true of other large investment shareholders, but the guidance of the United States is absolutely essential if these organizations are to perform their functions satisfactorily.

Several governments have been reviewing their policy towards the international satellite organizations in recent years, especially towards INTELSAT, which is by far the largest. In Europe, an EEC "Green Paper on the Development of the Common Market for Telecommunications Services and Equipment" concluded that it (the EEC) "has a long term interest in preserving major stabilizing factors of international satellite organizations" such as INTELSAT, INMARSAT and EUTELSAT. Of INTELSAT, the paper added "that INTELSAT currently faces two challenges: (1) adaptation of its organization and its strategy to a more competitive environment and (2) the international telecommunications community will have to develop new mechanisms for global coordination, including satellite and cable connections." Further development of INTELSAT, it added, "is largely linked to the evolution of transatlantic traffic and the interconnection issue regarding the newly emerging competitive U.S. service providers." The report also referred to the impact of fiber optic cables on the satellite services and revenues. The report did not mention that it is up to the European countries to adapt THEIR approach to INTELSAT and INMAR-SAT in order for the organizations to keep in line with present trends in user requirements. An increasingly common example of where this is not happening is that a country's policy toward INTELSAT/INMARSAT is expressed at an Assembly of Parties and/or a Meeting of Signatories, whereas its contribution to the organization's management is expressed by its monopoly telecommunications administration whose policy is sometimes based on profit and exclusion of competition. There are strong signs, however, that this situation is slowly changing and that a more active role is being taken by governments in the international telecommunications organizations.

In the United States the situation tends to be even more confused on account of the larger size of government and the larger number of interested government departments. During 1987, a Senate Committee on Commerce Science and Transportation, and a House of Representatives Energy and Commerce Subcommittee on Telecommunications held hearings and received responses which included the long-term policy towards INTELSAT and INMARSAT. These comments will eventually form a basis for determining future policy and the extent that changes may be needed to the Communications Satellite Act and the 1978 INMARSAT Act. This subject is large in itself, but is also interlinked with the United States' balance of trade problem (around 1986 to 1988) which attracts an even wider diversification of views. Relating purely to the satellite communications aspects, comments were diverse but on the whole supportive of INTELSAT and INMARSAT over their successes in the past and the prospects for the future. Unfortunately many comments inferred that INTELSAT and IN-MARSAT are United States organizations whose policy is wholly determined by the United States rather than cooperatively owned with other nations and in which the United States has a 25% share. It is sometimes forgotten that it is the two acts mentioned above that created the U.S. policy towards the organizations rather than creating the organizations themselves.

A former FCC Commissioner, The Hon. Abbott Washburn, who was largely responsible for United States' satellite policy during the late 1960s and early 1970s, urged the United States (talking of INTELSAT) not to "pull up the plant by the roots to see if it was healthy." He suggested instead that the United States "build on the foundation of the significant successes already achieved." He was critical of "loop-holes and leverage artists who are hard at work reinterpreting and in some cases circumventing the commitments and conditions set forth" in recent policy actions. These comments endorsed the fact that INTELSAT has a long history of being innovative and competitive.

At these hearings, the most intense criticism of INTELSAT came from a potential new entrant into the field of satellite communications, PanAmsat (see Chapter 3). This included a charge that INTELSAT once performed an important role, but its "past gold-plating has become a heavy burden to carry into this new world." Today, PanAmsat claims that "INTELSAT is ineffective and has done a particularly poor job of providing data and video services, which is one reason why PanAmsat can make money providing communications between the Americas. INTELSAT is a global monopolist that has deliberately set out to destroy PanAmsat and other separate systems by lowering prices and offering new services. INTELSAT has also abused Article XIV(d) of its Agreement, engaged in predatory pricing, and through its Signatories, [presumably the United States and some South American countries] purposefully boycotted separate systems by denying them landing rights." These and similar comments, ignored the facts that new services had been introduced by INTELSAT over many years, whereas PanAmsat had only been in existance since 1984, well after INTELSAT's introduction of domestic leased services (1975), leased international trans-ponder television services (1981), international business services (IBS) in 1983, and a thin-route VISTA service also in 1983. PanAmsat urged the United States government to "move quickly to end the adverse effects arising out of INTELSAT's monopoly." Most of PanAmsat's complaints would have been more appropriately launched against the United States' representation to INTELSAT rather than to INTELSAT itself; the interna-tional organization would be better employed using its resources to perform its own work rather than defending itself, at considerable expense, within the lobbies of its largest investor country.

Generally the comments of the various departments and those of major public telecommunication companies were supportive of INTELSAT. To-gether they reinforced the notion that the U.S. policy is not always clear, but what is worse for the rest of the world is that it is far from clear who motivates the actions of the United States signatory to the international organizations. Much of the future of satellite communications will depend upon government policies, especially those of the United States which is in most cases the largest user and investor.

10.1.2. Satellite and Launch Costs

Another very important factor for the future will be satellite and launch costs as these bear heavily upon any new or potential supplier of satellite services. Any new supplier must consider the cost of satellite manufacture, launching them safely, creating an operational control network, and obtain-ing sufficient revenue over a number of years to provide the kind of profit that is being sought.

There are many ways in which such financial calculations can be per-formed, and a vast number of different factors can be added to complicate the issue, but some very simple facts remain with figures that cannot be radically changed.

- Any reliable communications service must have a reliable contingency against failure. In this case, any satellite system must have a minimum of two satellites, although a sparing philosophy can alter as the number of satellites in the system increases.

- Each satellite needs to be launched into orbit. A two-satellite system cannot be defined as reliable if one satellite is on the ground and may take months to launch if the single orbiting satellite fails. Satellite launch services which were, until January 1986, only available from the United States, France, or the U.S.S.R., became almost unobtainable from the United States for a year or so, queued up in France, and commercially untried from the U.S.S.R. China is a new entrant into the launching field. Even if the Western launch situation improves before 1990, it will take some years before the backlog is disposed of unless many of the launches are cancelled. Until the Western launch prospects are normal again, the uncertainty will prevail. This uncertainty greatly increases the risk factor cost, and will remain a risk for several years until the launch market is opened again.

- Although launch insurance was not available for many years and any launch or satellite failure had to be faced as a straight financial loss, it is nevertheless considered advisable by even the largest operators such as INTELSAT. Some of the larger operators such as INTELSAT, INMARSAT, and EUTELSAT are studying the prospect of forming a mutual insurance organization. The loss of a new operator's first satellite and the vision of around US$150 million exploding in front of your eyes, or being lost in space, would be almost unimaginable without launch insurance. However, following the launch failures and space shuttle disaster in 1986, such insurance became hard, or impossible, to obtain. If obtainable at all, it became prohibitively expensive. Launch contractor incentives such as guaranteed relaunch or cash refund were considered as an alternative, but even this is increasing costs considerably.

- There are also operational costs. Ideally two satellites could be launched and all transponders sold or leased immediately, in which case the investor could walk away with the profit. This unfortunately is not the case; the satellites and all associated earth stations need to be controlled on a 24-hour a day basis for the life of the satellite system. This necessitates at least two earth stations for satellite control functions, tracking telemetry and command, and also for communications network control.

- Some kind of administrative organization, however basic, will be needed for these functions as well as legal, service billing, and any other functions that the company considers necessary. Even if the new competitor feels it possible to spend less than the existing satellite operators, cost will still be high; these costs are inherent in providing a reliable service.

Having briefly explained the need for some of the costs, let us now look at what the total costs may be for any new satellite communications operating company:

	US$ Million
Two satellites	120 to 250
Two launches	120 to 250
Two launch insurances	100
Pre-launch organization and legal costs	25
Two control earth stations	20
Ten years operating costs	Vary widely

The future of the launch industry is unknown but the implications for the western nations becoming reliant upon Soviet or Chinese launchers is not likely to be allowed to happen. The time that is passing while the launch problems are being resolved is providing the opportunity for operators, or potential operators, to reexamine whether or not their launch is really necessary or really financially worth the risk. My own view is that many will retire gracefully leaving the international organizations and large public telecommunications companies as the prime owners and operators.

10.1.3. Proliferation of Satellite Systems

There has never been, or ever likely to be, a "single global commerical telecommunications satellite system" as envisaged in the Preamble to the INTELSAT Agreement. The first signatories to that agreement took place in 1972; the U.S.S.R., which was not part of INTELSAT, and some of INTELSAT's first members were already planning their own systems. The single system has already become just a part (although still the largest part) of the world's satellite communications usage.

Many consider INTELSAT is afraid of competition, yet this has never been the case. INTELSAT, throughout its history, coordinated with new satellite systems. No new satellite system has even been rejected in the coordination process for competitive reasons; EUTELSAT, ARABSAT, Indonesia's Palapa, United States companies, European companies and many more have all been approved by INTELSAT. More will doubtless follow in the years to come. INTELSAT's cooperative structure should always be able to take as much competition that is offered, and in the course of time many of the existing nonprofitable systems could well revert to INTELSAT, provided that it remains a true cooperative structure, not too heavily dominated by any single country or group of countries.

An excess of satellite capacity will occur if a number of other competitive satellite systems emerge as well as the new fiber optic cable systems. This will be especially true in the North Atlantic and Pacific areas. This excess will likely be felt worse by the private satellite systems than the international organizations.

10.1.4. Fiber Optic Cables

To many, fiber optic cables are the death toll for satellites; to many others they are reason to say that satellite costs must be greatly reduced if satellite owners are to stay competitive. Both views are reminiscent of earlier years. In the 1920s many forecasters foresaw the end of cables with the advent of long distance radio; in the 1950s many forecasters foresaw the end of radio when submarine telephone cables appeared; in the 1960s, again forecasters said that cables would be superceded by satellites. Now it is time, apparently, for the satellites to give way to fiber optic cables!

Historically, and I see no reason for the situation to change, cables and wireless communications methods go side by side; each supports the other.

Fiber optic cables have enormous technological advantages over previous terrestrial transmission means, both wire and wireless. Because of their very high quality and capacity, they will provide their owners with the ability to reduce their rates, if they are forced to do so by competition. Domestically in the United States there are already several competing fiber optic cable systems, and as the quantity of spare unsold capacity increases, competition will become harder. The same situation is also likely to arise over the major international heavy traffic routes, the trans-North Atlantic and transpacific routes.

A study by the United States' Rand Corporation in 1986 said that of the large satellite operators, "COMSAT and INTELSAT face strong competition beginning in 1988 when the first fiber optic cable is scheduled to become operational." The report, among other things, recommended that the present uniform pricing policy and global cost averaging should be abandoned and that prices should be flexible on a route by route basis if INTELSAT is to survive competition with undersea fiber optic cables. This study is one of many, and each study is forced into making so many assumptions that the logic used can be turned into whatever direction was desired at the outset. The cost of transatlantic fiber cable is about US$350 million and it can be expected to last 25 years. For the same lifespan, two successive satellites would be necessary (assuming present lifespans) at a total cost of somewhat less than the cable.

A common pricing unit for the type of wideband system now being quoted commercially is the 1.5 Mbps T-1. INTELSAT's present annual charge for one end to a T-1 is under US$90,000. The Americal carrier AT&T's annual charge for one end of a T-1 via the TAT-8 fiber optic cable will be US$540,000. It is reasonable to deduct that about US$450,000 is for the terrestrial connection. If a fiber optic T-1 carrier were costed using reasonable loading, overhead, profit and other assumptions, it is likely that the cable cost would end up being very close to INTELSAT's present charge to its signatories. Most studies seem to agree that the break-even point, in terms of distance, is about 3000 km; above this satellites become advantageous. This is borne out, insofar as costs per circuit mile are concerned, in a 1985 NASA assessment of satellite communication trends in which the cost per circuit mile over about 2500 miles were:

Number of Voice Circuits	Cost (US$) Per Circuit Mile by	
	Satellite	*Cable*
8,000	1.5	6.25
20,000	1.3	2.46
32,000	1.25	1.51
65,000	0.90	0.90

The satellite still has the important advantage in that it can bypass the terrestrial carrier costs almost entirely by taking the service directly to, or very close to, the required destination with the same quality service. A few years ago this type of bypass was extremely difficult; most countries having only one satellite operator discouraged customers to own their own earth stations that would bypass existing terrestrial networks.

The largest international traffic stream is between the United States and the United Kingdom, presently about 3200 satellite circuits. These can be accommodated in one pair of satellite transponders. Under the present INTELSAT charging arrangements each voice circuit costs US$8880 a year on a cost per channel basis. If some flexibility is incorporated so that two transponders could be leased or purchased by the two countries at domestic service rates, then the actual circuit cost could drop to around US$1250 (leased) or US$312 (purchased). There is thus much flexibility available before satellites become uneconomic vis-à-vis fiber optic cables.

A more important factor, in my own view, is that the larger carriers on both sides of the ocean, and indeed all countries, do not want to trust all their treasures into the same egg basket. It is for this reason that the cables are necessary, not necessarily as an economic preference. This in itself may well spell some financial hard times for INTELSAT when traffic is transferred off the satellite initially to empty cables, but this is likely to be a temporary arrangement. It should also be borne in mind that the investors in the INTELSAT satellites, with the exception of the United States' signatory, are also investors in the cable systems, and they have no incentive for watching INTELSAT suffer financially. The United States' signatory, however, has more of a problem to the extent that it was created as a satellite corporation and until recently had no investment whatsoever in cable systems.

10.2. GROWTH IN PRESENT APPLICATIONS

INTELSAT's present percentage of the total international market is about 60%. This could drop if a lot of heavy route traffic moves from satellite to fiber optic cable, but there is no reason to believe that use of satellites for conventional telephone, data, and video applications is likely to drop, or become less profitable. Digital circuit multiplication techniques are already being introduced over the international satellite links which will keep the international organization economically competitive even in the era of fiber optic cables.

There is still a vast scope for growth in the area of private businesses as more and more businesses, both national and international, intercommunicate using computers. This need for digital communications is growing fast and satellites are particularly useful for such applications, especially where distances and numbers of locations become more widespread. Terrestrial routes, whether they be microwave or fiber optic cable, become increasingly expensive and increasingly vulnerable to failure as the network size increases, whereas satellite costs remain constant. The trend towards smaller earth stations continues and the cost of these small earth stations is also dropping considerably. The reliability of these earth stations also continues to improve and maintenance costs are dropping accordingly. In highly developed countries, fiber optic cable systems will attract many of these users due to shorter distances and competing networks, but internationally, except on the densest routes, satellites are likely to remain the only available option. The user requirements for this service, I believe, will increase dramatically over the next decade but there will probably be little firm forecasting on which the service providers will be able to plan ahead.

There is large scope for expansion in the field of international broadcasting and television. Growth may, however, be limited not so much by demand, but by the potential recipients being able, or prepared, to pay towards its transmission. A limited number of countries may copy the example of the United States and distribute its national programs for free reception by any country that choses to do so, but the number of countries likely to do this is small. An added impetus for Western nations to broadcast freely on a wider basis may come about when the U.S.S.R.'s TOR satellites start broadcasting on a worldwide basis in a few years time.

The advent of high definition television (HDTV), already being broadcast experimentally in Japan, will eventually increase the attractiveness of international as well as national television. However, more uniform standards in international television are urgently needed before this becomes practicable on a worldwide basis.

There is still a very high potential for growth in domestic satellite systems both in the number of countries using satellites and the capacity that they use. In view of the extremely high cost and risk of countries owning their own satellites, it would be economic common sense if some countries shared satellites. Regional satellites have not always proved either profitable or successful, and it would be sensible if more use was made of the existing INTELSAT system which would produce many cost advantages to these countries, as well as greatly reducing wastage of equatorial orbit resources. Some political pressures would probably be necessary to pursuade INTEL-SAT, or its signatories, that there would be great benefits for all if the organization improved its domestic service offerings by ensuring that all future transponders would be more suited to smaller earth stations and cheaper ground facilities.

Apart from future growth in the applications that are known to be profitable, it is also useful to look at where growth may emerge in applications that are not presently profitable, but which may well become profitable. There are very few thin-route networks yet using satellites and there remains an enormous potential for this type of application. Most of these networks would be domestic in nature as the international links are already available on a single circuit basis, but domestic thin-route networks would open up an entirely new market. The reason that this market has not developed is the cost of the earth station, and that in the majority of cases the revenue obtainable in return is often minimal; in other words, the national government must heavily subsidize this type of service, probably for many years. However, if any country is seriously considering national development, it must also be necessary to develop a simple communication structure first to the places and villages within its jurisdiction. This is a fact that is not always understood by the world's leading financial institutions such as the World Bank. Great strides are now being taken to reduce earth station costs and when this happens the demand for satellite capacity is likely to grow quickly and without warning. This would also open up a vast worldwide market for small earth stations, especially if they can be manufactured using local facilities in different countries.

Closely related to this type of thin-route network are educational and humanitarian applications for health and social improvement as well as ready-to-use networks for disasters and other emergencies. These applications seldom, if ever, produce revenues to the telecommunications administrations unless financed by government. There is nevertheless a great potential for this type of application if society could decide who would provide the financial support.

10.3. EMERGING FUTURE APPLICATIONS

10.3.1. The Advent of ISDN

The stated objective of telecommunications is now towards an integrated system digital network (ISDN) which will be entirely digital and include all services and applications. For this to be a truly global objective, satellites will, in the long-term, also be all digital for telephone, data, and video applications. This move towards digital services has already started and many services are already being converted to digital transmission quite rapidly. International demonstrations of ISDN have been taking place between the United States and the United Kingdom for a year or so and also between these and other countries. Although it will be several years before ISDN reaches many of the lesser developed countries where terrestrial networks are in poor condition, business and trade pressures could well hasten the process using satellite communications directly to customer premises.

If international trade is to continue to expand—which is the objective of most governments—other countries will be forced into the ISDN era if their businesses are to be able to communicate with other countries. In many countries the only way in which this looks like being achievable is by direct satellite access rather than using unreliable terrestrial networks. Although international business services have been introduced with the major trading countries first in mind initially, it must inevitably also move as a service to the smaller countries. This will generate new traffic and is unlikely to reduce the comparatively low existing international traffic in these countries. Many of these countries are also likely to turn more to satellites to provide the link reliability that is so necessary for any digital network. Although the fiber optic cables may provide advantages over satellites on a few selective routes, the transition to ISDN will create a much greater need for satellites for a wide variety of existing and new applications and services.

10.3.2. Direct Broadcasting Satellites

The broadcasting industry will shortly be able to decide whether direct broadcast satellites are likely to prove commercially viable or whether it is preferable to remain with lower cost satellites and already available equipment. My own view on this subject is that DBS satellites will not provide sufficient advantages over existing means to attract the high level of private investment necessary to offer a service profitably. However, where countries are subsidizing such satellites for industrial reasons, DBS satellites will be available anyway—whether they are profitable or not.

Whatever the outcome of the DBS satellites, there remains a vast potential to expand the worldwide international television services. During the last six years the number of full-time television services has grown substantially, mainly between the wealthier countries. Satellites are the only means of distributing worldwide television, and there are an increasing number of countries and organizations who are prepared to pay the cost for such distribution whether the recipients pay for receiving the service or not. The incentives for providing such distributions can be educational, political, religious, or several other reasons, but the attractiveness of the service remains, and is likely to provide considerable expansion potential for the next decade.

10.3.3. Mobile Applications

Lastly, the use of satellites for mobile applications is still a future growth potential for satellites. There is undoubtedly a great international need for providing more maritime mobile satellite services, introducing services to aircraft, and also possibly for ground vehicular services. The commercial viability of maritime mobile services has now been proved and these can be expected to expand and include a wider variety of safety services and expanded use to a larger number of smaller ships and pleasure craft.

Aeronautical mobile services can be expected to expand quickly as soon as they are introduced. Although available on a limited number of flights (mainly in the United States) they are urgently needed internationally, not only for the airlines and the aircraft themselves, but also for the passengers. Implementation of these services will be costly and will need to be supported, hopefully enough to defray costs, by the airline passengers using the service in the air. This, at the moment, is a totally unknown quantity. This new type of application is certainly not going to generate sufficient new revenues to warrant separate private satellites as well as INMARSAT. To this date the world's largest user, the United States, is still debating how such a service is to be organized after 20 years deliberation, and it will be a pity if the introduction of this important application is further delayed.

Mobile services to vehicular traffic, to my mind, is a service that may emerge and prosper during the 1990s. Early services will use the existing domestic satellites or those of the international organizations. If these early services prove successful, then the international organizations are likely to increase their payloads to match the new requirements.

10.4. CONCLUSION

I recall, in 1962, listening to a lecture by a very experienced leader in international commercial communications (Henry H. Eggers, then Managing Director of the Cable & Wireless Group) who said, approximately of the future ". . . the Americans are possibly going to introduce satellites for long distance communications. I doubt if they will affect us, but we shall follow their activities closely." It was a commercially cautious approach to the future, but his organization did follow satellite activities closely and turned out to be among the very first to use satellite communications commercially very much sooner than the speaker had imagined.

Telecommunications has always been an important industry; it has always been the thermometer for other industries. Much of the future will depend upon how much the satellite communications service industry is permitted to function on a cooperative basis rather than as a tool of individual governmental policies. Government policy should regulate each country's individual approach to the cooperative objective, not try to deregulate the international cooperatives themselves. Given that there is no stranglehold in the development of satellite communications imposed by governments, the future remains bright for both domestic and international satellite communications applications. One day, all phone calls by satellite may become "local calls".

We'll have sunrise in fifteen minutes. Meanwhile I'm rolling the ship so you can get a good view of the equatorial satellite belt. The brightest one—almost straight overhead—is INTELSAT's Atlantic-1 Antenna Farm. Then Intercosmos 2 to the West—that fainter star is Jupiter. And if you look just below that, you'll see a flashing light, moving against the star background—that's the new Chinese spacestation.

ALL THESE WORLDS ARE YOURS.
—2010: Odyssey Two, by Arthur C. Clarke. Copyright (c) 1984 by Serendib, B.V. By permission of the author and his agents, Scott Meredith Literary Agency, Inc., 845 3rd Ave., NY, NY 10022.

APPENDICES

INTERNATIONAL SATELLITE ORGANIZATIONS—MEMBERSHIPS AND PERCENTAGE SHARE[a]

COUNTRY	INTELSAT	INMARSAT	INTER-SPUTNIK	EUTELSAT	ARABSAT
Afghanistan	0.05		Member		
Algeria	0.25	0.05	User		0.90
Angola	0.11				
Antigua	User				
Argentina	1.00	0.12			
Australia	2.95	1.30			
Austria	0.33			1.96	
Bahamas	0.17				
Bahrain	User	0.07			4.00
Bangladesh	0.17				
Barbados	0.05				
Belgium	0.69	0.53		8.57	
Belize	User				
Benin	0.05				
Bolivia	0.12				
Botswana	User				
Brazil	1.29	1.23			
Bulgaria		0.24	Member		
Burkina Faso	0.05				
Burundi	User				
Cameroon	0.24				
Canada	2.75	2.17			
Cape Verde	User				

[a] INTELSAT percentages as at March 1988; INMATSAT percentages as at December 1987; EUTELSAT percentages as at October 1987; ARABSAT percentages as at December 1986.

283

COUNTRY	INTELSAT	INMARSAT	INTER-SPUTNIK	EUTELSAT	ARABSAT
Central African Rep.	0.05				
Chile	0.57	0.05			
China, People's Rep.	1.54	0.41			
Colombia	1.18	0.05			
Congo	0.05				
Cook Islands	User				
Costa Rica	0.05				
Cote d'Ivoire	0.26				
Cuba			Member		
Cyprus	0.19			0.97	
Czechoslovakia			Member		
Denmark	0.49	1.48		0.53	
Djibouti	User				YES
Dominican Rep.	0.10				
Ecuador	0.35				
Egypt	0.70	0.12			10.40
El Salvador	0.05				
Ethiopia	0.07				
Fiji	0.05				
Finland	0.16	0.20		0.24	
France	4.21	2.50		14.85	
Gabon	0.05	0.06			
Gambia	User				
Germany, Federal D.R	3.61	1.79		11.51	
Germany, People's D.R		0.05	Member		
Ghana	0.05				
Greece	0.68	1.99		0.67	
Guatamala	0.05				
Guinea	0.05				
Guyana	User				
Haiti	0.16				
Hondurus	0.05				
Hungary			Member		
Iceland	0.18			0.05	
India	1.22	0.44			
Indonesia	0.38	0.38			
Iran	0.72	0.07			
Iraq	0.30	0.05	Member		2.60
Ireland	0.13			0.05	
Israel	0.66	0.05	User		
Italy	2.24	1.43		7.93	
Jamaica	0.38				
Japan	4.70	8.65			
Jordan	0.30				3.30

COUNTRY	INTELSAT	INMARSAT	INTER-SPUTNIK	EUTELSAT	ARABSAT
Kenya	0.40				
Kiribati	User				
Korea, Republic of	1.32	0.35			
Korea, People's Rep.	User		Member		
Kuwait	0.77	0.63			8.20
Laos			Member		
Lebanon	0.13				6.30
Leotho	User				
Liberia	User	0.13			
Libya	0.15		User		18.2
Liechtenstein	0.05			0.05	
Luxembourg	0.05			4.98	
Madagascar	0.05				
Malawi	0.13				
Malaysia	0.70	0.07			
Maldives	User				
Mali	0.09				
Malta	User			0.05	
Mauritius	0.14				
Mauritania	0.05				0.20
Mexico	0.60				
Monaco	0.05			0.05	
Mongolia			Member		
Morocco	0.17				0.50
Mozambique	User				
Nepal	Member				
Netherlands	1.11	2.30		9.00	
New Zealand	0.72	0.10			
Nicaragua	0.05		Member		
Niger	0.05				
Nigeria	0.97				
Norway	0.33	15.93		4.65	
Oman	0.31	0.05			0.60
Pakistan	0.39	0.05			
Palestine					0.20
Panama	0.05	0.05			
Papua New Guinea	0.05				
Paraguay	0.11				
Peru	0.56	0.05			
Philippines	0.64	0.12			
Poland	User	0.33	Member		
Portugal	0.50	0.09		1.32	
Qatar	0.25	0.05			5.00
Romania	User		Member		
Rwanda	0.05				

COUNTRY	INTELSAT	INMARSAT	INTER-SPUTNIK	EUTELSAT	ARABSAT
Sao Tome E Principe	User				
Saudi Arabia	1.54	0.83			26.60
San Marino				0.05	
Senegal	0.05				
Seychelles	User				
Singapore	1.20	2.36			
Solomon Islands	User				
Somalia	0.05				0.30
South Africa	1.15				
Spain	1.84	1.75		1.00	
Sri Lanka	0.05	0.06			
Sudan	0.18				2.10
Suriname	User				
Swaziland	0.05				
Sweden	0.65	0.62		1.66	
Switzerland	1.09			9.40	
Syria	0.12		User		1.70
Tanzania	0.05				
Tchad	0.05				
Thailand	0.55				
Togo	0.05				
Tonga	User				
Trinidad and Tobago	0.05				
Tunisia	0.05	0.05			0.60
Turkey	0.23			0.93	
Uganda	0.05				
U. Arab Emirates	1.36	0.10			6.00
United Kingdom	14.21	17.59		20.00	
United States	26.42	28.56	User		
Uruguay	0.05				
USSR	User	2.34	Member		
Vanuatu	User				
Vatican City	0.05			0.05	
Venezeula	1.00				
Viet Nam	0.05		Member		
Western Samoa	User				
Yemen, Arab Rep.	0.11				0.70
Yemem, Peoples	User		Member		0.60
Yugoslavia	0.20			0.05	
Zaire	0.24				
Zambia	0.13				
Zimbabwe	Member				
Total membership	117	51	16	26	22

SATELLITE SERIES
CHARACTERISTICS SUMMARY

This appendix provides some of the basic characteristics of present-day satellite series, and some of the series being planned for the near future. The purpose is to provide the reader with an approximate comparison between the satellite series

Coverage and EIRP data are approximate only. Names of beam coverages differ widely. Global beams provide the widest possible coverage to the whole globe as seen from the satellite. Zonal beams include hemispheric, conus, and wide area beams. Spot beams are narrow beams pointed at much smaller areas of the world. The power (EIRP) varies considerably depending on where an earth station is located within the beam's coverage. The figures quoted should not be used for calculation purposes but are approximate for edge of beam.

Costs are also approximate. Some purchasers add incentive payments to the basic price which, where known, are included. Costs, given in US$, are the prices when purchased at the approximate exchange rate at the time.

International Service Satellites

	INTELSAT V	INTELSAT V-A	INTELSAT VI	INTELSAT VII	INMARSAT II	INTERSPUTNIK GOROZONT	INTERSPUTNIK TOR	USA PanAmSat
Number in series	9	6	5	5	3–6	13	14	1–2
First launch	1980	1985	1989	1992	1989	1978	1990	1988
Satellite contractor	Ford	Ford	GM/Hughes	Ford	BR. Aerospace	U.S.S.R.	U.S.S.R.	GE/RCA
Design life (years)	7	7	10	10	10	3–5	3–5	11
Stabilization	3-axis	3-axis	Spin	3-axis	3-axis	3-axis	3-axis	3-axis
Circuit capacity	12,500	12,500	100,000					
E.O.L. power supply (W)	1,290	1,290	2,250	3,500	1,200			1,400
Dimensions								
Mass at start of life (kg)	2,140	2,140	2,531	457	1,160	954		1,273
Mass at end of life (kg)	950	950	1,878		690			693
Width (m)	2.0	2.0	3.6		1.48	2		1.9
Height (m)	6.4	6.4	11.8		2.5	5		
Span (m)	15.7	15.7	N/A		15.23	10.5		15.8
Communications								
Total Bandwidth (MHz)	2,137	2,322	3,230	2,408		204	1440	1440
L-band					18		Ka	–
C-band (MHz)	1,357	1,542	2,344	1,532	23	204	–	864
Ku-band (MHz)	780	780	886	876		36	–	432
Ka/Q-band (GHz)							3	–
Transponders								
C-band	32	32	46	36	9	6		
Bandwidth (MHz)	26	26	36	16 × 72 MHz / 10 × 36 MHz		34		12 × 36 MHz / 6 × 72 MHz
Polarization	72	72	72	Circular	Circular	Circular		Linear
TWT output power (W)	Circular	Circular	Circular	10/16	4.5	15/40		8.5/16.2
Ku-band	4.5	4.5/8.5	5.5			1		6
Bandwidth (MHz)	6	6	10	6 × 72 MHz / 4 × 112 MHz		36		72
Polarization	72/77/241	72/77/241	72/150	Linear		Linear		Linear
TWT output power (W)	Linear	Linear	Linear	20–50				16.2
L-band	10	10	18.5		1			
Coverage and EIRP								
C-band global (dBW)	23.5	23.5	26.5	26.6–29		28–35		26–35
zonal (dBW)	26–29	26–29	31	33		33		34–39
spot (dBW)	–	34		33		48		
Ku-band spot (dBW)	40–43	40–43	42–45	42–47		38.5		34–49
Approximate cost per satellite, excluding launch and insurance (US$m)	35	60	140	100	60			155

288

Regional Satellites

	EUTELSAT I	EUTELSAT II	ARABSAT
Number in series	5	4–8	2
First launch	1984	1990	1985
Satellite contrator	Br. Aerospace	Aerospatiale	Aerospatiale
Design life (years)	7	10	7
Stabilization	3-axis	3-axis	3-axis
E.O.L. power supply (W)	1000	3200	1400
Dimensions			
Mass at start of life (kg)	1100	1625	1270
Mass at end of life (kg)	610	620	592
Width (m)	2.5	2.5	2.26
Height (m)	2.0	2.0	1.64
Span (m)	13.8	22.4	20.7
Communications			
Total bandwidth (MHz)	864	1152	858
C-band (MHz)	–	–	858
Ku-band (MHz)	864	1152	–
Transponders			
C-band	–	–	26
Bandwidth (MHz)	–	–	33
Polarization	–	–	Circular
TWT output power (W)	–	–	8.5
Ku-band	12–14	16	–
Bandwidth (MHz)	72	72	–
Polarization	Linear	Linear	–
TWT output power (W)	20	50	–
Coverage and EIRP			
C-band zonal (dBW)	–	–	31
spot (dBW)	–	–	–
Ku-band spot (dBW)	38–42	45–51	–
Approximate cost per satellite, excluding launch and insurance (US$m)	50	80	68

Domestic Satellite systems

	Australia AUSSAT I	Brazil BRASILSAT	Canada ANIK C	Canada ANIK D	France TELECOM I	India INSAT I	Indonesia PALAPA B	Mexico MORELOS	Germany TV-SAT	Germany Kopernikus	Italy Italsat	Japan CS
Number in series	3	2	3	2	3	3	3	2	2	2	1	3
First launch	1985	1985	1985	1982	1984	1982	1983	1985	1987	1989	1990	1983
Satellite contractor	GM/Hughes	Spar, Canada	Spar, Canada	Spar, Canada	MATRA	Ford	GM/Hughes	GM/Hughes	EUROSAT	Siemens	Selenia	Ford-NEC
Design life (years)	10	10	10	8	7	7	10	10	7	10	5	5
Stabilization	Spin	Spin	Spin	Spin	3-axis	3-axis	Spin	Spin	3-axis	Spin	3-axis	3-axis
E.O.L. power supply (W)	1,074	800	900	1,000	1,100	1,200	940	940	3,300	1,700	1,560	840
Dimensions												
Mass at start of life (kg)	1,250	1,140	1,140	1,240	1,185	1,090	1,200	1,200	1,180	1,400	1,700	550
Mass at end of life (kg)	650	671	567	660	685	620	635	660	900	815	890	
Width (m)	2.16	2.16	2.16	2.16	2.18	1.5	1.9	2.1	1.6			2.3
Height (m)	7.09	7.09	7.09	7.09	2.96	2.1	3.2	3.2	2.4			3.2
Span (m)					16	5			18			
Communications												
Total bandwidth (MHz)	675	864	1,080	864	536	432	864	1,584	135			500
C-band (MHz)	–	864	–	864	320	432	864	936	–			500
Ku-band (MHz)	675	–	1,080	–	216	–	–	648	135			–
Ka-band (GHz)	–	–	–	–			–	–	–		1800	1
Transponders												
C-band	–	24	–	24	4	12	24	20	–	–	2	2
Bandwidth (MHz)		36		36	40/120	36	36	36/72			180	180
Polarization		Linear		Linear	Circular	Circular	Linear	Linear			?	?
TWT output power (W)		9.6		11.5	8.5	?	10	7-10			?	?
Ku-band	15				6			6	5	15	6 Ka	10 Ka-
Bandwidth (MHz)	45				36			108	27	44/90	36/120	130
Polarization	Linear		Linear		Linear			Linear	Linear	Linear	Linear	Linear
TWT output power (W)	12/30		15		20			20	230	10/20	20	5
Coverage and EIRP												
C-band global (dBW)		38			35	28.5	32	33				29.5
zonal (dBW)					34.5							
spot (dBW)		48	38						40	50		
Ku-band spot (dBW)	36–43		48		44–47			45	63	48		
Ka-band spot (dBW)											47/57	37
Approximate cost per satellite, excluding launch and insurance (US$m)	60	50	40	40	80	40	40	46	100	130	150	220

Domestic Satellite Systems *(Continued)*

	USA AT&T TELESTAR 3	USA GTE GSTAR	USA GTE SPACENET	USA Hughes GALAXY I-IV	USA GM/Hughes GALAXY VVII	USA CONTEL AMSAT	USA GE/RCA SATCOM 4-7	USA GE/RCA SATCOM K	USA MIC SBS1-4	USA GM/Hughes WESTAR4-6
Number in series	3	4	4	4	3	3	4	2	4	3
First launch	1983	1986	1984	1983	1989	1985	1982	1985	1980	1982
Satellite contractor	Gm/Hughes	GE/RCA	GE/RCA	GM/Hughes	GM/Hughes	GE/RCA	GE/RCA	GE/RCA	GM/Hughes	GM/Hughes
Design life (years)	10	10	10	10	10	10	10	10	7	10
Stabilizaton	Spin	3-axis	3-axis	Spin	Spin	3-axis	3-axis	3-axis	Spin	Spin
E.O.L. power supply (W)	670	1,400	1,200	2,240	1,425	1,215	800	1,395	900	700
Dimensions										
Mass at start of life (kg)	1,200	1,400	1,200	1,200	1,350	1,255	1,080	1,395	1,060	
Mass at end of life (kg)	655	760	705	654	666	800	?		585	
Width (m)	2.7	1.9	1.0	3.6	3.6	1.3	1.3	1.7	2.1	2.1
Height (m)	11.8	1.9	1.3	11.8	11.8	1.6	1.6	2.2	6.5	6.5
Span (m)	–	16	14.5	–	–	1.3	1.3	2.2	–	–
Communications										
Total bandwidth (MHz)	864	1,188	1,296	864	832	1,260	864	1,080	430	864
C-band (MHz)	864	–	864	864	–	756	864	–	–	864
Ku-band (MHz)	–	1,188	432	–	832	504	–	1,080	430	–
Transponders										
C-band	24	–	12 + 6	24	–	21	24	–	–	24
Bandwidth (MHz)	36	–	36–72	36	–	36	36	–	–	36
Polarization	Linear	–	Linear	Linear	–	Linear	Linear	–	–	Linear
Output power (W)	5.5	–	8.5	8.5	–	8.5	8.5	–	–	7.5
Ku-band	–	22	6–	–	16	7	–	20	10	–
Bandwidth (MHz)	–	54	72	–	52	72	–	54	43	–
Polarization	–	Linear	Linear	–	Linear	Linear	–	Linear	Linear	–
Output power (W)	–	20/30	16	–	50	16.5	–	50	20	–
Coverage and EIRP										
C-band global (dBW)	–	–	–	28	–	–	27	–	–	27
conus/zonal (dBW)	34	–	33–38	34	–	35	34	–	–	34
Hawaii spot (dBW)	34	–	–	28	–	–	26	–	–	28
Ku-band spot (dBW)	–	40–45	40–44	–	50	34–42	–	50	42–47	–
Approximate cost per satellite, excluding launch and insurance (US$m)	45	35	75	40	100	40	75	80	80	40

291

Domestic and Regional TV Broadcast Satellites

	France TDF-1	Luxembourg ASTRA/GDL	Sweden TELE-X	China ASIASAT	Japan BS-2	Japan JSCC "Superbird"	Japan JCSAT	UK BSB
Number in series		2		2	2	2	2	2
First launch	1989	1988	1989	1989	1984	1990	1989	1989
Satellite contractor	EUROSAT	GE/RCA	EUROSAT	GM/Hughes	Toshiba	Ford	GM/Hughes	GM/Hughes
Design life (years)	9	10	10	10		13	10	10
Stabilization	3-axis	3-axis	3-axis	Spin	3-axis	3-axis	Spin	Spin
E.O.L. power supply (W)	3,000	3,576	3,400	700	800	4,000	2,200	1,000
Dimensions								
Mass at start of life (kg)	2,080	1,820	2,130		670	1,350	1,350	1,250
Mass at end of life (kg)	1,030	1,045	1,200		350			700
Width (m)	1.6		1.6		1.3	1.8	3.6	
Height (m)	2.4		2.4		2.9	1.8	11.8	
Span (m)	18		18		9	24	–	
Communications								
Total Bandwidth (MHz)	825	250	400	864		720	500	800
C-band (MHz)	–	–	–	864	–	–	–	–
Ku-band (MHz)	825	250	400	–		720	500	800
Transponders								
C-band				24				
Bandwidth (MHz)				36				
Polarization				Linear				
TWT output power (W)				10				
Ku-band	5	16	4		2	18	32	16
Bandwidth (MHz)	27	15	27/40/86			36	27	27
Polarization	Linear	Linear	Linear			Linear	Linear	Circular
TWT output power (W)	230	230	50		100	20	20	
Coverage and EIRP								
C-band global (dBW)	–	–	–	27	–	–	–	–
zonal (dBW)	–	–	–	34	–	–	–	–
spot (dBW)	–	–	60–65	28	–	–	–	–
Ku-band spot (dBW)	64	50		–	–	49	50	64
Approximte cost per satellite, excluding launch and insurance (US$m)	100	50	100	120	48	160	152	200

Domestic Systems

	China Chinasat	Israel Amos	Iran Zohren	Iraq Babylonsat	Ireland Atlantic
Number in series		2	1–3	2	
First launch	1988	1992	1992	1993	1991
Satellite contractor	China	Israel			GM/Hughes
Design life (years)	10				10
Stabilization	3-axis				Spin
E.O.L. power supply (W)	1,000				1,000
Dimensions					
Mass at start of life (kg)	1,630				1,200
Mass at end of life (kg)	908				700
Width (m)					
Height (m)					
Span (m)					
Communications					
Total bandwidth (MHz)	200				200
C-band (MHz)	200				200
Ku-band (MHz)	–				–
Transponders					
C-band	16				
Bandwidth (MHz)	36				
Polarization	Linear				
TWT output power (W)	10/20				
Ku-band					16
Bandwidth (MHz)	–				27/54
Polarization	–				Linear
TWT output power (W)	–				10/50
Coverages and EIRP					
C-band global (dBW)	35				
zonal (dBW)					
spot (dBW)	45				
Ku-band spot (dBW)	–	48	45		40/55
Approximate cost per satellite, excluding launch and insurance (US$m)					150

SATELLITE LAUNCHES

Satellite	Flight	Owner	Date (mm/dd/yy)	Launcher	Location
ACTS		USA/NASA	12/31/90	Shuttle	Planed 101°W
AMS	F-1	Israel	12/31/93	Ariane	Planned 15°E
AMSAT	F-1	USA/Contel	08/27/85	Shuttle	128°W
AMSAT	F-2	USA/Contel	06/17/88	Ariane	81°W
AMSAT	F-3	USA/Contel	12/31/90		Planned
ANIK A	F-1	Canada/Telesat	11/09/72	Delta	Retired
ANIK A	F-2	Canada/Telesat	04/20/73	Delta	Retired
ANIK A	F-3	Canada/Telesat	05/07/75	Delta	Retired
ANIK B	F-1	Canada/Telesat	12/15/78	Delta	Retired
ANIK C	F-1	Canada/Telesat	04/12/85	Shuttle	Sold to Asiasat
ANIK C	F-2	Canada/Telesat	06/18/83	Shuttle	110°W
ANIK C	F-3	Canada/Telesat	11/11/82	Shuttle	117°W
ANIK D	F-1	Canada/Telesat	08/27/82	Delta	105.5°W
ANIK D	F-2	Canada/Telesat	11/08/84	Shuttle	11.5°W
ANIK E	F-1	Canada/Telesat	12/31/89	Delta	Planned 104.5°W
ANIK E	F-2	Canada/Telesat	12/31/90		Planned 117.5°W
ARABSAT	F-1	ARABSAT	02/08/84	Ariane	19°E
ARABSAT	F-2	ARABSAT	06/17/84	Shuttle	26.5°E
ASIASAT	F-1	UK/Asiasat	12/31/89	L-March	Planned
ASIASAT	F-2	UK/Asiasat	12/31/90	L-March	Planned
ASTRA	F-1	Luxembourg	12/10/88	Ariane	19°E
ASTRA	F-2	Luxembourg/SES	12/31/90	Ariane	Planned 1°E
ATS	F-1	USA/NASA	12/07/66	Atlas	Retired
ATS	F-2	USA/NASA	04/06/67	Atlas	Retired
ATS	F-3	USA/NASA	11/08/67	Atlas	Retired
ATS	F-4	USA/NASA	08/10/67	Atlas	Retired
ATS	F-5	USA/NASA	08/12/69	Atlas	Retired
ATS	F-6	USA/NASA	05/30/74	Titan	Retired
AURORA	F-1	USA/Alascom	10/27/82	Delta	146°W
AURORA	F-2	USA/Alascom	12/31/90		Planned
AURORA	F-3	USA/Alascom	12/31/91		Planned
AUSSAT A	F-1	Australia	08/15/85	Shuttle	156°E
AUSSAT A	F-2	Australia	11/15/85	Shuttle	160°E
AUSSAT A	F-3	Australia	09/16/87	Ariane	164°E
BABYLONSAT	F-1	Iraq	04/30/93		Planned 30°
BRAZILSAT	F-1	Brazil	02/08/85	Ariane	65°W
BRAZILSAT	F-2	Brazil	03/28/86	Ariane	72°W
BS-2	A	Japan/NHK	01/23/84	N2	110°E
BS-2	B	Japan/NHK	02/12/86	N2	109°E

Satellite	Flight	Owner	Date (mm/dd/yy)	Launcher	Location
BSB	F-1	UK/BSB	12/31/89	Delta	Planned 329°E
BSB	F-2	UK/BSB	06/30/90		Planned 329°E
COMSTAR	F-1	USA/Comgen	05/13/76	Atlas	Retired
COMSTAR	F-2	USA/Comgen	07/22/76	Atlas	Retired
COMSTAR	F-3	USA/Comgen	06/29/78	Atlas	Retired
COMSTAR	F-4	USA/Comgen	02/21/79	Atlas	76°W
COSMOS		USSR	05/17/82	Proton	76°E
CS-2	A	Japan/NTT	02/04/83	N2	131°E
CS-2	B	Japan/NTT	09/16/88	N2	130°E
CS-3	A	Japan/NTT	12/31/88	H1	Planned
CS-3	B	Japan/NTT	12/31/88	H1	Planned
China	F-15	China	04/08/84		125°E
Chinasat	F-1	China	12/31/88	L-March	Planned
Chinasat	F-2	China	12/31/91	L-March	Planned
DFS	F-1	Germany	05/06/89	Ariane	23.5°E
DFS	F-2	Germany	11/30/89	Ariane	Planned 28.5°E
DFS	F-3	Germany	12/31/90	Ariane	Planned spare
ELKRAN	F-1	USSR	10/27/76	Proton	Retired
ELKRAN	F-10	USSR	03/12/83	Proton	Retired
ELKRAN	F-11	USSR	09/30/83	Proton	47.5°E
ELKRAN	F-12	USSR	03/16/84	Proton	Retired
ELKRAN	F-13	USSR	08/24/84	Proton	107°E
ELKRAN	F-14	USSR	03/22/85	Proton	99°E
ELKRAN	F-15	USSR	05/24/86	Proton	98°E
ELKRAN	F-2	USSR	09/20/77	Proton	Retired
ELKRAN	F-3	USSR	02/21/79	Proton	Retired
ELKRAN	F-4	USSR	10/03/79	Proton	Retired
ELKRAN	F-5	USSR	07/14/80	Proton	Retired
ELKRAN	F-6	USSR	12/26/80	Proton	Retired
ELKRAN	F-7	USSR	06/26/81	Proton	Retired
ELKRAN	F-8	USSR	02/05/82	Proton	Retired
ELKRAN	F-9	USSR	09/16/82	Proton	65°E
ETS	F-2	Japan	02/23/77	N	Retired 1978
ETS	F-5	Japan	08/27/87	H	150°E
ETS	F-6	Japan	12/31/92	H-2	Planned
EUTELSAT I	F-1	EUTELSAT	06/01/84	Ariane	16°E
EUTLESAT I	F-2	EUTELSAT	08/04/84	Ariane	7°E
EUTELSAT I	F-3	EUTELSAT	01/01/85	Ariane	Launch failure
EUTELSAT I	F-4	EUTELSAT	09/16/87	Ariane	13°E
EUTELSAT I	F-5	EUTELSAT	07/21/88	Ariane	350°E
EUTELSAT II	F-1	EUTELSAT	12/31/90	Ariane	Planned 12°E
EUTELSAT II	F-2	EUTELSAT	12/31/90	Ariane	Planned 3°E
EUTELSAT II	F-3	EUTELSAT	12/31/90	Ariane	Planned 36°E
FLTSATCOM	F-1	USA/Military	02/09/78	Atlas	177°W
FLTSATCOM	F-2	USA/Military	05/04/79	Atlas	Retired
FLTSATCOM	F-3	USA/Military	01/18/80	Atlas	23°W
FLTSATCOM	F-4	USA/Military	10/31/80	Atlas	172°E
FLTSATCOM	F-5	USA/Military	08/06/81	Atlas	Failed
FLTSATCOM	F-6	USA/Military	03/27/87	Atlas	Failed
FLTSATCOM	F-7	USA/Military	04/12/86	Atlas	105°W
FLTSATCOM	F-8	USA/Military	12/31/88	Atlas	Retired
GOES	F-1	USA/NASA	10/16/75	Delta	Retired
GOES	F-2	USA/NATO	06/16/77	Delta	Retired
GOES	F-3	USA/NASA	06/16/78	Delta	Failed
GOES	F-4	USA/NASA	09/08/80	Delta	45°W
GOES	F-5	USA/NASA	05/22/81	Delta	67°W
GOES	F-6	USA/NASA	04/28/83	Delta	135°W
GOES	F-7	USA/NASA	05/31/86	Delta	Failed

Satellite	Flight	Owner	Date (mm/dd/yy)	Launcher	Location
GOES	F-8	USA/NASA	02/26/87	Delta	75°W
GORIZONT	F-1	USSR	12/19/78	Proton	Retired
GORIZONT	F-10	USSR	08/01/84	Proton	79°E
GORIZONT	F-11	USSR	01/18/85	Proton	140°E
GORIZONT	F-12	USSR	06/10/86	Proton	15°W
GORIZONT	F-13	USSR	11/18/86	Proton	90°E
GORIZONT	F-2	USSR	07/05/79	Proton	Retired
GORIZONT	F-3	USSR	12/28/79	Proton	Retired
GORIZONT	F-4	USSR	06/14/80	Proton	11°W
GORIZONT	F-5	USSR	03/15/82	Proton	Retired
GORIZONT	F-6	USSR	10/20/82	Proton	Retired
GORIZONT	F-7	USSR	06/30/83	Proton	349°E
GORIZONT	F-8	USSR	11/30/83	Proton	89°E
GORIZONT	F-9	USSR	04/22/84	Proton	97°E
GStar	F-1	USA/GTESpacenet	05/08/85	Ariane	105°W
GStar	F-2	USA/GTESpacenet	03/28/86	Ariane	103°W
GStar	F-3	USA/GTESpacenet	09/08/88	Ariane	Failed
Galaxy	F-1	USA/GM-Hughes	06/28/83	Delta	133.5°W
Galaxy	F-2	USA/GM-Hughes	09/22/83	Delta	74°W
Galaxy	F-3	USA/GM-Hughes	09/21/84	Delta	93°W
Galaxy	F-4	USA/GM-Hughes	06/30/91		Planned 91°W
Galaxy	F-5	USA/GM-Hughes	10/31/91		–
Galaxy	F-6	USA/GM-Hughes	06/30/90		Planned 91°W
IMNARSAT II	F-1	INMARSAT	12/31/89	Delta	Planned
INMARSAT II	F-2	INMARSAT	12/31/89	Delta	Planned
INMARSAT III	F-1	INMARSAT	12/31/94		Planned
INSAT I	F-1	India/Insat	04/10/82	Delta	Failed in orbit
INSAT I	F-2	India/Insat	08/31/83	Shuttle	74°E
INSAT I	F-3	India	07/21/88	Ariane	Failed in orbit
INSAT I	F-4	India/Insat	02/28/89	Delta	Planned
INTELSAT I	F-1	INTELSAT	04/06/65	Delta	Retired
INTELSAT II	F-1	INTELSAT	10/26/66	Delta	Launch failure
INTELSAT II	F-2	INTELSAT	01/11/67	Delta	Retired
INTELSAT II	F-3	INTELSAT	03/22/67	Delta	Retired
INTELSAT II	F-4	INTELSAT	09/27/67	Delta	Retired
INTELSAT III	F-1	INTELSAT	09/18/68	Delta	Launch failure
INTELSAT III	F-2	INTELSAT	12/18/69	Delta	Retired
INTELSAT III	F-3	INTELSAT	02/05/69	Delta	Retired
INTELSAT III	F-4	INTELSAT	05/21/69	Delta	Retired
INTELSAT III	F-5	INTELSAT	07/25/70	Delta	Launch failure
INTELSAT III	F-6	INTELSAT	01/14/70	Delta	Retired
INTELSAT III	F-7	INTELSAT	04/22/70	Delta	Retired
INTELSAT III	F-8	INTELSAT	07/23/70	Delta	Launch failure
INTELSAT IV	F-1	INTELSAT	05/22/75	Atlas	Retired
INTELSAT IV	F-2	INTELSAT	01/25/71	Atlas	Retired
INTELSAT IV	F-3	INTELSAT	12/19/71	Atlas	Retired
INTELSAT IV	F-4	INTELSAT	01/23/72	Atlas	Retired
INTELSAT IV	F-5	INTELSAT	06/13/72	Atlas	Retired
INTELSAT IV	F-6	INTELSAT	02/20/75	Atlas	Launch failure
INTELSAT IV	F-7	INTELSAT	08/23/73	Atlas	Retired
INTELSAT IV	F-8	INTELSAT	11/21/74	Atlas	Retired
INTELSAT IVA	F-1	INTELSAT	09/25/75	Atlas	Retired
INTELSAT IVA	F-2	INTELSAT	01/29/76	Atlas	Retired
INTELSAT IVA	F-3	INTELSAT	01/06/78	Atlas	Retired
INTELSAT IVA	F-4	INTELSAT	05/26/77	Atlas	Retired
INTELSAT IVA	F-5	INTELSAT	09/29/77	Atlas	Launch failure
INTELSAT IVA	F-6	INTELSAT	03/31/78	Atlas	Retired
INTELSAT V	F-1	INTELSAT	05/24/81	Atlas	174°E

Satellite	Flight	Owner	Date (mm/dd/yy)	Launcher	Location
INTELSAT V	F-2	INTELSAT	12/06/80	Atlas	359°E
INTELSAT V	F-3	INTELSAT	12/15/81	Atlas	307°E
INTELSAT V	F-4	INTELSAT	03/04/82	Atlas	325.5°E
INTELSAT V	F-5M	INTELSAT	09/28/82	Atlas	63°E
INTELSAT V	F-6M	INTELSAT	05/19/83	Atlas	342°E
INTELSAT V	F-7M	INTELSAT	10/18/83	Atlas	66°E
INTELSAT V	F-8M	INTELSAT	05/04/84	Atlas	180°E
INTELSAT V	F-9M	INTELSAT	06/09/84	Atlas	Launch failure
INTELSAT VA	F-10	INTELSAT	03/22/85	Atlas	335.5°E
INTELSAT VA	F-11	INTELSAT	06/29/85	Atlas	332.5°E
INTELSAT VA	F-12	INTELSAT	09/28/85	Atlas	60°E
INTELSAT Va	F-13	INTELSAT	05/17/88	Ariane	307°E
INTELSAT VA	F-14	INTELSAT	05/30/86	Ariane	Launch failure
INTELSAT VA	F-15	INTELSAT	01/26/89	Ariane	60°E
INTELSAT VI	F-1	INTELSAT	09/30/89	Ariane	Planned
INTELSAT VI	F-2	INTELSAT	11/30/89	Titan	Planned
INTELSAT VI	F-3	INTELSAT	01/31/90	Titan	Planned
INTELSAT VI	F-4	INTELSAT	12/31/90	Ariane	Planned
INTELSAT VI	F-5	INTELSAT	12/31/91	Ariane	Planned
INTELSAT VII	F-1	INTELSAT	12/31/92		Planned
INTELSAT VII	F-2	INTELSAT	12/31/92		Planned
INTELSAT VII	F-3	INTELSAT	12/31/93		Planned
INTELSAT VII	F-4	INTELSAT	12/31/93		Planned
INTELSAT VII	F-5	INTELSAT	12/31/93		Planned
ITALSAT		Italy	12/31/90	Ariane	Planned
JCSAT	F-1	Japan/JCSAT	03/06/89	Ariane	150°E
JCSAT	F-2	Japan/JCSAT	12/31/89	Ariane	Planned
JECS	F-1	Japan	02/06/79	N	Failed 1980
LEASAT	F-1	USA/Hughes	11/08/84	Atlas	Failed 1985
LEASAT	F-2	USA/Hughes	08/30/84	Atlas	15°W
LEASAT	F-3	USA/Hughes	04/12/85	Atlas	Failed
LEASAT	F-4	USA/Hughes	08/29/85	Atlas	Failed
LEASAT	F-5	USA/Hughes			Unlaunched
LINCOLN	F-1	USA/MIT	03/14/76	Titan	Retired
MARECS	F-1	Europe/ESA	21/20/81	Ariane	178°E
MARECS	F-2	Europe/ESA	09/15/82	Ariane	Launch failure
MARECS	F-3	Europe/ESA	11/10/84	Ariane	334°E
MARISAT	F-1	USA/Comgen	02/19/76	Delta	345°E
MARISAT	F-2	USA/Comgen	06/09/76	Delta	73°E
MARISAT	F-3	USA/Comgen	10/14/76	Delta	176°E
METEOSAT	F-1	Europe/ESA	01/15/88	Ariane	17°W
MOLNIYA I	F-1	USSR	04/23/65	Vostok	Nonsynchronous
MOLINYA I	F-70	USSR	12/26/86	Vostok	Nonsynchronous
MOLNIYA II	F-1	USSR	11/24/71	Vostok	Nonsynchronous
MOLNIYA II	F-17	USSR	02/11/77	Vostok	Nonsynchronous
MOLINYA III	F-1	USSR	11/21/74	Vostok	Nonsynchronous
MOLINYA III	F-30	USSR	10/21/86	Vostok	Nonsynchronous
MORELOS	F-1	Mexico	06/17/85	Shuttle	113°W
MORELOS	F-2	Mexico	11/26/85	Shuttle	116°W
NATO II	F-1	NATO	03/20/70	Delta	Retired
NATO II	F-2	NATO	02/03/71	Delta	Retired
NATO III	F-1	NATO	04/22/76	Delta	30.5°W
NATO III	F-2	NATO	01/28/77	Delta	60°W
NATO III	F-3	NATO	11/19/78	Delta	18°W
OLYMPUS		Europe/ESA	04/30/88	Ariane	
OTS	F-1	Europe/ESA	09/13/77	Delta	Launch failure
OTS	F-2	Europe/ESA	05/11/78	Delta	4°E
PANAMSAT	F-1	USA/PanAmSat	07/17/88	Ariane	45°W

Satellite	Flight	Owner	Date (mm/dd/yy)	Launcher	Location
PANAMSAT	F-2	USA/PanAmSat	12/31/90		Planned
Palapa A	F-1	Indonesia	07/08/76	Delta	Retired
Palapa A	F-2	Indonesia	10/03/77	Delta	Retired
Palapa B	F-1	Indonesia	06/18/83	Shuttle	108°E
Palapa B	F-2	Indonesia	02/03/84	Shuttle	Lost, recovered
Palapa B	F-3	Indonesia	12/31/88	Delta	Planned 118°E
RADUGA	F-1	USSR	12/22/75	Proton	Retired
RADUGA	F-10	USSR	10/09/81	Proton	81°E
RADUGA	F-11	USSR	11/26/82	Proton	34°E
RADUGA	F-12	USSR	04/08/83	Proton	77°E
RADUGA	F-13	USSR	08/26/83	Proton	158°E
RADUGA	F-14	USSR	02/15/84	Proton	75°E
RADUGA	F-15	USSR	06/22/84	Proton	129°E
RADUGA	F-16	USSR	08/08/85	Proton	47°E
RADUGA	F-17	USSR	11/15/85	Proton	34°E
RADUGA	F-18	USSR	01/17/86	Proton	25°W
RADUGA	F-19	USSR	10/25/86	Proton	44°E
RADUGA	F-2	USSR	09/11/76	Proton	Retired
RADUGA	F-3	USSR	07/24/77	Proton	Retired
RADUGA	F-4	USSR	07/18/78	Proton	Retired
RADUGA	F-6	USSR	04/25/79	Proton	Retired
RADUGA	F-6	USSR	02/20/80	Proton	Retired
RADUGA	F-7	USSR	10/05/80	Proton	Retired
RADUGA	F-8	USSR	03/18/80	Proton	Retired
RADUGA	F-9	USSR	07/21/81	Proton	Retired
RELAY	F-1	USA/NASA	12/13/62	Delta	Retired
RELAY	F-2	USA/NASA	01/21/64	Delta	Retired
SATCOM	F-1	USA/RCA	12/12/75	Delta	Retired
SATCOM	F-1R	USA/RCA	04/30/83	Delta	136°W
SATCOM	F-2	USA/RCA	03/16/79	Delta	Retired
SATCOM	F-2R	USA/RCA	09/08/83	Delta	72°W
SATCOM	F-3	USA/RCA	12/06/79	Delta	Launch failure
SATCOM	F-3R	USA/RCA	11/19/81	Delta	131°W
SATCOM	F-4	USA/RCA	01/15/82	Delta	83°W
SBS	F-1	USA/MCI	11/15/80	Delta	97°W
SBS	F-2	USA/MCI	09/24/81	Delta	95°W
SBS	F-3	USA/MCI	11/11/82	Shuttle	97°W
SBS	F-4	USA/MCI	08/30/84	Shuttle	91.5°W
SBS	F-5	USA/MCI	09/08/88	Ariane	101°W
SIRIO	F-1	Italy	08/25/77	Delta	Retired
SKYNET I	F-1	UK/Military	11/21/69	Delta	Failed 1970
SKYNET I	F-2	UK/Military	08/19/70	Delta	Failed
SKYNET II	F-1	UK/Military	01/19/74	Delta	Failed
SKYNET II	F-2	UK/Military	11/23/74	Delta	130°E
SKYNET IV	F-1	UK/Military		Shuttle	Planned
SKYNET IV	F-2	UK/Military	12/10/88	Shuttle	
SKYNET IV	F-3	UK/Military		Shuttle	Planned
SPACENET	F-1	USA/GTE Spacenet	05/22/84	Ariane	120°W
SPACENET	F-2	USA/GTE Spacenet	11/09/84	Ariane	70.5°W
SPACENET	F-3	USA/GTE Spacenet	09/11/85	Ariane	Launch failure
SPACENET	F-3R	USA/GTE Spacenet	03/11/88	Ariane	87°W
SPOTNET	F-1	USA/Nat. Exch.	06/30/89	Ariane	Planned
SPUTNIK	I	USSR	10/04/57		Retired
STW	F-1	China	04/03/84		126°E
STW	F-2	China	02/01/86		103°E
STW	F-3	China	03/07/88		87°E
SYNCOM	F-1	USA/RCA	02/13/63	Delta	Failed in orbit
SYNCOM	F-3	USA/RCA	07/26/63	Delta	Retired

Satellite	Flight	Owner	Date (mm/dd/yy)	Launcher	Location
SYNCOM	F-3	USA/RCA	08/19/64	Delta	Retired
SYNCOM	F-4	USA/RCA	08/30/84	Delta	177°W
Superbird	F-1	Japan/JSCC	03/31/89	Titan	Planned
Superbird	F-2	Japan/JSCC	10/31/89	Titan	Planned
Symphonie	F-1	France	12/18/74	Delta	Retired
Symphonie	F-2	France	08/27/75	Delta	Retired
TDF	F-1	France	10/27/88	Ariane	19°W
TDF	F-2	France	11/30/89	Ariane	Planned 19°W
TDRSS	F-1	USA/NASA	04/04/83	Shuttle	42°W
TDRSS	F-2	USA/NASA	01/15/86	Shuttle	Lost
TDRSS	F-3	USA/NASA	09/28/88	Shuttle	171°W
TDRSS	F-4	USA/NASA	03/13/89	Shuttle	
TELECOM I	A	France	08/04/84	Ariane	8°E
TELECOM I	B	France	05/08/85	Ariane	Failed 1987
TELECOM I	C	France	03/11/88	Araine	5°W
TELSTAR I	F-1	USA/AT&T	07/10/62	Delta	Retired
TELECOM II	F-1	France	01/31/92	Ariane	Planned 352°E
TELECOM II	F-2	France	06/30/92	Ariane	Planned 355°E
TELECOM II	F-3	France		Ariane	Planned 3°E
TELSTAR III	301	USA/AT&T	07/28/83	Atlas	96°W
TELSTAR III	302	USA/AT&T	08/30/84	Shuttle	85°W
TELSTAR III	303	USA/AT&T	06/19/85	Shuttle	125°W
TVSAT	F-1	Germany	11/21/87	Ariane	Failed in orbit
TVSAT	F-2	Germany	05/31/89	Ariane	Planned 19°W
Tele-X	F-1	Sweden	04/02/89	Ariane	5°E
TONGASAT	F-1	Tonga	11/30/91		Planned 105.5°E
TONGASAT	F-2	Tonga	12/31/92		Planned 115.5°E
TONGASAT	F-3	Tonga			Planned 121.5°E
TONGASAT	F-4	Tonga			Planned 131°E
TONGASAT	F-5	Tonga			Planned 187.5°E
TONGASAT	F-6	Tonga			Planned 170.5°E
TONGASAT	F-7	Tonga			Planned 164°E
TONGASAT	F-8	Tonga			Planned 160°E
WESTAR	I	USA/GM-Hughes	04/13/74	Delta	Retired
WESTAR	II	USA/GM-Hughes	10/10/74	Delta	Retired
WESTAR	III	USA/GM-Hughes	08/10/79	Delta	92°W
WESTAR	IV	USA/GM-Hughes	02/26/82	Delta	99°W
WESTAR	V	USA/GM-Hughes	06/09/82	Delta	122°W
WESTAR	VI	USA/GM-Hughes	02/03/84	Shuttle	Sold to Asiasat
ZOHREH	F-1	Iran	04/30/91		Planned 34°E

WORLDWIDE SATELLITE TELEVISION CHANNELS

Degrees East	Satellite	TV Program	Originating Country	Band	Power (dBW)
1	ASTRA	Starts 1989	Luxembourg	Ku	50
5	TELECOM	Canal J	France	Ku	44
		La Cinq	France	Ku	44
		TV-6	France	Ku	44
		Cristal	France	Ku	44
7	EUTELSAT	WorldNet	USA	Ku	40
		Eurovision	EBU	Ku	40
		Pace		Ku	40
10	EUTELSAT	NRK	Norway	Ku	40
		RA1	Italy	Ku	40
		TVE1	Spain	Ku	40
13	EUTELSAT	Filmnet	Belgium	Ku	40
		TV5	France	Ku	40
		3Sat	Germany	Ku	40
		Satl	Germany	Ku	40
		RA1	Italy	Ku	40
		RTL Plus	Luxembourg	Ku	40
		Teleclub	Switzerland	Ku	40
		Sky Channel	UK	Ku	40
		Super	UK	Ku	40
		Worldnet	USA	Ku	40
		Arts		Ku	40
		Filmnet	Belgium	Ku	40
19	ARABSAT			C	21
53	GORIZONT	Programma 1	USSR	C	46
		Programma 2	USSR	C	46
60	INTELSAT	3Sat	Germany	Ku	42
		Eins Plus	Germany	Ku	42
		Eureka	Germany	Ku	42
		Music Box	Germany	Ku	42
		WDR3	Germany	Ku	42
		BR3	Germany	Ku	42
		AFRTS	Germany	Ku	42
		2 Occasional use channels		C	18

301

Degrees East	Satellite	TV Program	Originating Country	Band	Power (dBW)
63	INTELSAT	National TV	Iran	Ku	42
		National TV	Iran	Ku	42
		National TV	France	C	26
		National TV	South Africa	C	26
		National TV	India	C	26
		2 Occasional use channels		C	18
66	INTELSAT	National TV	Algeria	C	18
		National TV	Ethiopia	C	18
		National TV	Malaysia	C	18
		National TV	Zaire	C	18
		National TV	China	C	26
		National TV	Turkey	Ku	42
		2 Occasional use channels		C	18
74	INSAT	National TV	India	S	
77	PALAPA	QTR7	Indonesia	C	33
		RPN9	Philippines	C	33
		RTM2	Malaysia	C	33
83	INSAT	National TV	India	S	
90	GORIZONT	Programma 1	USSR	C	46
		Programma 2	USSR	C	46
108	PALAPA	Army TV	Indonesia	C	33
		TVR1	Indonesia	C	33
		RTM1	Malaysia	C	33
		TBS7	Thailand	C	33
110	BS-2	NHK	Japan	Ku	48
156	AUSSAT			Ku	36–45
160	AUSSAT				
164	AUSSAT				
174	INTELSAT	2 Occasional use channels		C	18
180	INTELSAT	Channel 7	Australia/USA	C	23
		Channel 9	Australia/USA	C	23
		Channel 10	Australia/USA	C	23
		AFRTS	USA	C	26
		NHK	Japan/USA	C	23
		JISO	Japan/USA	C	18
		National TV	France	C	26
		National TV	New Zealand	C	26
		NZBC	New Zealand/Can	C	26
		2 Occasional use channels		C	18
214	AUORA	Alaska TV	USA	C	32
221	RCA SATCOM	Nicholodeon	USA	C	26
		PTL	USA	C	26
		Trinity	USA	C	26
		Financial News	USA	C	26
		Viewers Choice	USA	C	26
		SPN	USA	C	26
		ESPN	USA	C	28

Degrees East	Satellite	TV Program	Originating Country	Band	Power (dBW)
		CBN/QVC	USA	C	26
		Cable West	USA	C	26
		Showtime West	USA	C	26
		MTV	USA	C	28
		Eternal Word	USA	C	26
		HBO	USA	C	26
		Open Net	USA	C	26
		Video Hits	USA	C	28
		Travel	USA	C	26
		Lifetime	USA	C	26
		Reuters	USA	C	26
		Nat. Jewish	USA	C	26
		BET	USA	C	26
		Weather	USA	C	26
		Information	USA	C	26
		Home Shopping	USA	C	26
		Cinemax West	USA	C	26
		A&E	USA	C	26
226	GALAXY	TNN Nashville	USA	C	28–34
		CBN Cable	USA	C	29–35
		Disney	USA	C	29–34
		Showtime East	USA	C	29–35
		S.I.N.	USA	C	28–34
		CNN	USA	C	29–35
		ESPN	USA	C	29–35
		Movie Channel	USA	C	28–34
		C-Span	USA	C	29–35
		WOR, New York	USA	C	29–34
		WTBS, Atlanta	USA	C	28–34
		Cinemax	USA	C	28–34
		Galavision	USA	C	28–34
		USA Cable	USA	C	29–35
		Discovery	USA	C	28–34
		HBO	USA	C	28–34
		Disney	USA	C	28–34
		Senate	USA	C	28–35
		Christian	USA	C	25
		Baptist BTN	USA	C	25
		Occasional use channels		C	25
242.5	ANIK	Knowledge	Canada	Ku	48
		Access	Canada	Ku	48
		Premier	Canada	Ku	48
		Superchannel	Canada	Ku	48
		Atlantic ASN	Canada	Ku	48
		La Sette	Canada	Ku	48
		TV Ontario	Canada	Ku	48
		Radio Quebec	Canada	Ku	48
247	MORELOS	XEW	Mexico	Ku	45
		XHDF	Mexico	Ku	45
		XHTM	Mexico	Ku	45

Degrees East	Satellite	TV Program	Originating Country	Band	Power (dBW)
255.5	ANIK	TSN Sports	Canada	C	35
		Much Music	Canada	C	35
		Global	Canada	C	35
		CHCH Hamilton	Canada	C	35
		CBC English	Canada	C	33
		CBC French	Canada	C	33
		TCTV Toronto	Canada	C	33
		CITV Edmonton	Canada	C	33
		Parliament English		C	33
		Parliament French		C	33
		CBC Montreal	Canada	C	33
		BCTVVancouver	Canada	C	33
		Hollywood	USA	C	33
		PBS Detroit	USA	C	33
		CBS Detroit	USA	C	33
		Occasional use channels		C	33
257	GSTAR	Technology	USA	Ku	31
		Campus	USA	Ku	34
		ESPN	USA	Ku	34
		CNN Headlines	USA	Ku	31
		Showtime East	USA	Ku	32
		Showtime West	USA	Ku	36
		BuzNet	USA	Ku	32
261	WESTAR	World TV	USA	C	30
		World of Faith	USA	C	30
		PBS (4 Chans)	USA	C	30
		Occasional use channels		C	30
264	TELSTAR	CBS Central	USA	C	34
		CBS Regional	USA	C	34
		CBS National	USA	C	34
		ABC Regional	USA	C	34
		ABC Regional	USA	C	34
		ABC Occasional	USA	C	34
		ABC East/West	USA	C	34
		World TV	USA	C	34
		Independent	USA	C	34
		Occasional use channels		C	34
		NBC Central	USA	Ku	34
		NBC West	USA	Ku	34
		NBC Affiliates	USA	Ku	34
		PSN Network	USA	Ku	34
274	TELSTAR	ABC Pacific	USA	C	34
		ABC	USA	C	34
		CBS East/West	USA	C	34
		CBS Pacific	USA	C	34
		CBS Regular	USA	C	34
		CBS Toronto	Canada	C	34

Degrees East	Satellite	TV Program	Originating Country	Band	Power (dBW)
		NETCOM	USA	C	34
		Occasional use channels		C	34
279	SATCOM	Home Shopping	USA	C	28
		Bravo	USA	C	28
		Commerce	USA	C	28
		Nickolodeon	USA	C	28
		ABC Affiliates	USA	C	28
		Madison Square	USA	C	28
		Na. Childrens	USA	C	28
		NETCOM	USA	C	28
		Sports Vision	USA	C	28
		Movie Classics	USA	C	28
		Home Sports	USA	C	28
		Playboy	USA	C	28
		New England Sports		C	28
		Silent Network	USA	C	30
		Hit Video	USA	C	30
		WPIX New York	USA	C	30
		Prime Ticket	USA	C	30
		Nostalgia	USA	C	30
		Nightline	USA	C	30
		NASA Contract	USA	C	30
307		National TV	Peru	C	26
		Sydney Ch. 9	USA/UK	C	23
		CBS/TVE	Canada/Spain	C	23
315	PANAMSAT				
325.5	INTELSAT	TVE	Spain	C	18
332.5	INTELSAT	National TV	Chile	C	23
		National TV	Nigeria	C	18
		National TV	Colombia	C	18
		National TV	Sudan	C	18
		National TV	Libya	C	18
		National TV	Venezuela	C	23
		ABC	USA/UK	C/Ku	
		CBS	USA/UK	C/Ku	
		NBC	USA/UK	C/Ku	
		CNN	USA/UK	C/Ku	
		Brightstar	USA/UK	C/Ku	
		Eurovision	USA/Italy	C/Ku	
		Worldnet	USA/Brazil	C	18
		W. Smith	UK	Ku	40
		BBC	UK	Ku	40
		MTV	UK	Ku	40
		Financial News	UK	Ku	40
		Scansat	UK	Ku	40
		Premier	UK	Ku	40
		DGT	France	C	26
		Childrens	UK	Ku	40
		Kindernet		Ku	40

Degrees East	Satellite	TV Program	Originating Country	Band	Power (dBW)
		Lifestyle	UK	Ku	40
		Screensport	UK	Ku	40
		Information	UK	Ku	40
349	GORIZONT	Moskva TV			
356	GORIZONT	Moskva TV			
359	INTELSAT	AFRTS	USA	C	26
		New World	Norway	Ku	40
		SVT-1	Norway	Ku	40
		SVT-2	Norway	Ku	40
		National TV	Israel	Ku	40

SATELLITES IN THE GEOSYNCHRONOUS ORBIT
(including planned)

Degrees East	Country	System Name	Frequency Bands				Degrees West
			L	C	Ku	Ka	
1.0	Luxembourg	Astra			Ku		359.0
3.0	France	Telecom		C	Ku		357.0
5.0	Sweden	Tele-X			Ku		355.0
7.0	France	EUTELSAT			Ku		353.0
8.0	USSR	See Note 1	L	C			352.0
10.0	France	EUTELSAT			Ku		350.0
13.0	Italy	ITALSAT				Ka	347.0
14.0	Nigeria	Future		C			346.0
15.0	Israel	Future		C	Ku		345.0
16.0	Italy	Future			Ku		344.0
17.0	Saudi Arabia	SABS			Ku		343.0
19.0	Luxembourg	Astra		C	Ku		341.0
19.0	Italy	Future			Ku		338.0
20.0	Nigeria	Future		C			340.0
23.0	USSR	Statsionar	L	C			337.0
23.0	Germany	DFS			Ku	Ka	337.0
26.0	Iran	Future			Ku		334.0
26.0	Saudi Arabia	ARABSAT		C			334.0
27.0	USSR	TOR				Ka	333.0
28.5	Germany	DFS			Ku	Ka	331.5
30.0	Iraq	Babylonsat			Ku		330.0
32.0	USSR	TOR				Ka	328.0
32.0	France	TDF			Ku		328.0
34.0	Iran	Future			Ku		326.0
35.0	USSR	See Note 1	L	C			325.0
38.0	Pakistan	Future			Ku		322.0
40.0	USSR	Statsionar		C			320.0
41.0	Iran	Future			Ku		319.0
41.0	Pakistan	Future			Ku		319.0
45.0	USSR	See Note 2	L				315.0

Degrees East	Country	System Name	L	C	Ku	Ka	Degrees West
47.0	Iran	Future		C	Ku		313.0
53.0	USSR	See Note 2	L	C	Ku		307.0
57.0	INTELSAT	Indian 1		C	Ku		303.0
60.0	INTELSAT	India 2		C	Ku		300.0
62.0	USSR	TOR				Ka	298.0
63.0	INTELSAT	Indian 3		C	Ku		297.0
64.5	INMARSAR		L	C	Ku		296.5
65.0	USSR	TOR				Ka	295.0
65.0	Italy	Sirio			Ku		295.0
66.0	INTELSAT	Indian 4		C	Ku		294.0
73.0	France	MARECS	L	C			287.0
74.0	Indian	Insat	L	C			286.0
76.0	USSR	GOMSS	L			Ka	284.0
77.0	USSR	CSSRD			Ku		283.0
80.0	USSR	Intersputnik		C			280.0
81.5	USSR	Foton		C			278.5
83.0	India	Insat	L	C			277.0
85.0	USSR	See Note 2	L	C	Ku		275.0
87.5	China	Chinasat		C			272.5
90.0	USSR	See Note 2	L	C	Ku		270.0
93.5	India	Insat	L	C			266.5
95.0	USSR	See Note 4		C	Ku		265.0
98.0	China	Chinasat		C			262.0
99.0	USSR	Statsionar		C			261.0
102.0	India	Iscom		C			258.0
103.0	USSR	See Note 2	L	C	Ku		257.0
105.5	UK	Asiasat		C			254.5
105.5	Tonga	Tongasat		C			254.5
108.0	Indonesia	Palapa		C			252.0
110.0	Japan	BS			Ku		250.0
113.0	Indonesia	Palapa		C			247.0
115.5	Tonga	Tongasat		C			244.5
116.0	UK	Asiasat		C			244.0
116.0	Indonesia	Palapa		C			242.0
121.5	Tonga	Tongasat		C			238.5
122.0	UK	Asiasat		C			238.0
125.0	China	STW		C			235.0
128.0	USSR	See Note 1	L	C			238.0
130.0	USR	Statsionar		C			230.0
131.0	Tonga	Tongasat		C			229.0
132.0	Japan	CS		C		Ka	228.0
140.0	USSR	See Note 1	L	C	Ku		220.0
140.0	Japan	Meteorology	L				220.0
145.0	USSR	Statsionar		C			215.0
150.0	Japan	JCSAT		C	Ku	Ka	210.0
154.0	Japan	JCSAT		C	Ku	Ka	206.0
156.0	Australia	Aussat			Ku		204.0
158.0	Japan	Superbird			Ku		202.0
160.0	Australia	Aussat			Ku		200.0
160.0	Tonga	Tongasat		C			200.0
162.0	Japan	Superbird			Ku		198.0

Degrees East	Country	System Name	L	C	Ku	Ka	Degrees West
164.0	Australia	Aussat			Ku		196.0
164.0	Tonga	Tongasat		C			196.0
166.0	USSR	GOMS	L				194.0
167.0	Papua N.G.	Pacstar		C	Ku		193.0
170.5	Tonga	Tongasat		C			189.5
174.0	INTELSAT	Pacific 1		C	Ku		186.0
176.5	USA	Marisat	L	C			183.5
177.0	INTELSAT	Pacific 2		C	Ku		183.0
180.0	INTELSAT	Pacific 3		C	Ku		180.0
182.0	USA	Finansat		C			178.0
185.0	Papua N.G.	Pacstar		C	Ku		175.0
187.5	Tonga	Tongasat		C			172.5
189.0	USA	TDRS				Ka	171.0
190.0	USSR	See Note 2	L	C	Ku		171.0
191.5	USSR	Foton		C			169.5
192.0	USSR	Potok		C			168.0
200.0	USSR	ESDRN			Ku		160.0
214.0	USA	Aurora		C			146.0
215.0	Mexico	Ilhuicahua		C	Ku		145.0
215.0	USSR	Volna	L				145.0
216.0	USA	Westar		C			144.0
217.0	USA	Aurora		C			143.0
218.5	Mexico	Ilhuicahua		C	Ku		141.5
220.0	USA	Galaxy		C			140.0
221.0	USA	RCA Satcom		C			139.0
223.0	USA	Future		C			137.0
224.0	USA	GStar			Ku		136.0
226.0	USA	Galaxy		C			134.0
227.0	USA	Westar			Ku		133.0
228.0	USA	Comstar		C	Ku		132.0
229.0	USA	RCA Satcom		C			131.0
230.0	USA	Future		C	Ku		130.0
232.0	USA	Amsat		C			128.0
223.0	USA	Comstar		C			127.0
234.0	USA	Future			Ku		126.0
235.0	USA	Telstar		C	Ku		125.0
236.0	USA	Fedex			Ku		124.0
236.5	USA	Westar		C			123.5
237.5	USA	Westar		C			122.5
240.0	USA	Spacenet		C	Ku		120.0
241.0	USA	Dominion		C	Ku		119.0
241.0	USA	RCA Satcom		C			119.0
242.5	Canada	Anik			Ku		117.5
243.5	Mexico	Morelos			Ku		112.5
246.5	Mexico	Morelos			Ku		116.5
247.5	Canada	Anik			Ku		113.5
248.5	Canada	Anik		C			111.5
250.0	Canada	Anik			Ku		110.0
251.5	Canada	Anik		C	Ku		108.5
252.5	Canada	Anik			Ku		107.5
253.5	Canada	Musat	L				106.5

Degrees East	Country	System Name	L	C	Ku	Ka	Degrees West
255.0	USA	GStar			Ku		105.0
255.5	Canada	Anik		C	Ku		104.5
257.0	USA	GStar			Ku		103.0
259.0	USA	SBS			Ku		101.0
261.0	USA	SBS			Ku		99.0
261.0	USA	Westar		C			99.0
263.0	USA	SBS			Ku		97.0
263.0	Cuba	Future		C			97.0
264.0	USA	Telstar		C			96.0
265.0	USA	SBS			Ku		95.0
266.0	USA	Future		C	Ku		94.0
267.0	USA	Fordsat		C	Ku		93.0
269.0	USA	Westar		C			91.0
271.0	Ecuador	Future		C			89.0
271.5	USA	Spacenet		C			88.5
273.0	USA	Comstar		C			87.0
274.0	USA	Telstar		C			86.0
275.0	USA	RCA Satcom			Ku		85.0
275.0	Argentian	Future		C	Ku		85.0
277.0	Cuba	Future		C			83.0
277.0	USA	Amsat		C	Ku		83.0
279.0	USA	RCA Satcom		C	Ku		81.0
280.0	Argentina	Future		C	Ku		80.0
281.0	USA	TDRSS				Ka	79.0
282.5	Ecuador	Future		C			77.5
283.0	USA	Fedex			Ku		77.0
284.0	USA	Comstar		C			76.0
284.0	Colombia	Future		C			76.0
285.0	Colombia	Future		C			75.0
285.0	USA	Comstar			Ku		75.0
286.0	USA	Galaxy		C			74.0
288.0	USA	Westar		C			72.0
288.0	Ecuador	Future		C			72.0
289.0	USA	Future			Ku		71.0
290.0	Brazil	Brazilsat		C			70.0
291.0	USA	Spacenet		C	Ku		69.0
293.0	USA	RCA Satcom		C	Ku		67.0
294.0	USA	Future		C	Ku		66.0
295.0	Brazil	Brazilsat		C			65.0
296.0	USA	Future			Ku		64.0
297.0	USA	Future		C			63.0
298.0	USA	SBS			Ku		62.0
299.0	USA	Future		C			61.0
300.0	USA	Future		C	Ku		60.0
302.0	USA	ISI		C			58.0
303.0	USA	Panamsat		C	Ku		57.0
304.0	INTELSAT	Future		C	Ku		56.0
304.0	USA	Future		C	Ku		56.0
305.0	USA	Future		C	Ku		55.0
307.0	INTELSAT	Atlantic 1		C	Ku		53.0
310.0	INTELSAT	Atlantic 2		C	Ku		50.0

Degrees East	Country	System Name	Frequency Bands				Degrees West
			L	C	Ku	Ka	
310.0	USA	Future			Ku		50.0
313.0	USA	Finansat		C	Ku		47.0
315.0	USA	Panamsat		C	Ku		45.0
316.0	France	Videosat			Ku		44.0
317.0	USA	Future		C	Ku		43.0
319.0	USA	Future		C	Ku		41.0
319.5	INTELSAT	Future		C	Ku		40.5
322.5	France	Videosat			Ku		37.5
322.5	USA	Orion			Ku		37.5
325.5	INTELSAT	Atlantic 3		C	Ku		34.5
329.0	INTELSAT	Atlantic 4		C	Ku		31.0
329.0	UK	BSB			Ku		31.0
329.0	Ireland	Eiresat			Ku		31.0
332.5	INTELSAT	Atlantic 5		C	Ku		27.5
332.5	USSR	Statsionar		C			27.5
333.5	USST	Volna	L				26.5
334.0	INMARSAT	Atlantic 1	L	C			26.0
335.0	USSR	See Note 1	L	C	Ku		25.0
335.5	INTELSAT	Atlantic 6		C	Ku		24.5
337.0	France	Marecs		C			23.0
338.0	INTELSAT	Atlantic 7		C	Ku		21.5
340.0	Luxembourg	Astra		C	Ku		20.0
341.0	Switzerland	Future			Ku		19.0
341.0	France	TDF			Ku	Ka	19.0
341.0	Luxembourg	Future			Ku		19.0
341.0	Italy	Future			Ku		19.0
341.0	Germany	TVSAT			Ku		19.0
342.0	INTELSAT	Atlantic 8		C	Ku		18.0
344.0	USSR				Ku		16.0
345.0	INMARSAT	Atlantic 2		C			15.0
345.0	USSR	Foton/Potok		C			15.0
346.0	USSR	Intersputnik	L	C	Ku		14.0
349.0	France	Future			Ku	Ka	11.0
351.0	USSR	Statsionar		C			8.5
352.0	France	Telecom		C	Ku		8.0
355.0	France	Telecom		C	Ku		5.0
359.0	INTELSAT	Atlantic 9		C	Ku		1.0

Note 1: Volna (L) and Statsionar (C)
Note 2: Volnar, Statsionar, and Loutch (K)
Note 3: Statsionar and Potok (L)
Note 4: Statsionar and CSDRN (KU)
Military Satellites are not included in this list.

APPENDIX F

BIBLIOGRAPHY

Bagleghole, K. C., *A Century of Service*, The Anchor Press Ltd., Tiptree, Essex, U.K., 1969.

Barnett, S. K., Botein, M. and Noam, E. M., *Law of International Telecommunications in the United States*, Nomos Veriagsegesellschaft, Baden-Baden, West Germany, 1988.

Barty-King, H., *Girdle Round The Earth*, Heinemann, London, U.K., 1979.

Colino, R. R., *The INTELSAT Definitive Arrangements—Ushering in a New Era*, European Broadcasting Union, 1973.

Colino, R. R., Commentary: The Case AGAINST Deregulating the Global Satellite Network, *International Journal of Satellite Communications*, Vol. 3, 1985.

Colino, R. R., A Chronicle of Policy and Procedure: The Formulation of a Reagan Administration Policy on International Satellite Telecommunications, *Journal of Space Law*, Vol. 13, No. 2, 1985.

Crispin, J., Cummins, J. H., Lemus, R. and Reyna, J., Satellite Versus Fibre Optic Cables, *International Journal of Satellite Communications*, March, 1985.

Eward, R., *The Deregularization of International Telecommunications*, Artech House, Dedham, MA, 1987.

Harcourt, E., *Taming the Tyrant—The First 100 Years of Australian International Communications*, Allen & Unwin, Sydney, Australia, 1987.

Jane's Spaceflight Directory, Jane's Publishing, London, U.K., 1985.

Johnson, L. L., Excess Capacity in International Telecommunications, *Telecommunications Policy*, September, 1987.

Kildow, J. T., *INTELSAT: Policy-maker's Dilemma*, Lexington Books, Lexington, MA, 1973.

Lee, Y.-S., Competition between fiber optic cables and satellites on transatlantic routes, *ITU Telecommunications Journal*, Vol. 54, XII, 1987.

Lloyd's, *Satellite Types and Prices Handbook: 1985*, Lloyd's, London, U.K., 1985.

Lloyd's, *World Satellite Survey: A Guide to Commercial Satellite Communications*, Lloyd's Aviation Department, London, U.K., 1988.

Longman, *International Directory of Telecommunications*, Longman Press, New York, 1984.

Markey, D. J., Commentary: The Case FOR Deregulating the Global Satellite Network, *International Journal of Satellite Communications*, Vol. 3, pp. 246–246, 1985.

Martin, D. L., *Communications Satellites 1958–1988*, The Aerospace Corporation, El Segundo, CA, 1988.

Martin, J., *Communications Satellite Systems*, Prentice Hall, Englewood Cliffs, NJ, 1978.

Martinez, L., *Communications Satellites: Power Politics in Space*, Artech House, Dedham, MA, 1985.

McWhinney, E. (Ed.), *The International Law of Communications*, Dobbs Ferry, New York, 1971.

Miya, K., *Satellite Communications Engineering*, KDD, Tokyo, Japan, 1985.

Morgan, W. L., *Business Earth Stations for Telecommunications*, Wiley Series in Telecommunications, Wiley-Interscience, New York, 1988.

Pelton, J. N., *Global Telecommunications Satellite Policy*, Lomond Books, Mount Airy, MD, 1974.

Pelton, J. N. and Alper, J. R. (Eds.), The INTELSAT Global Satellite System, *Progress in Astronautics & Aeronautics*, Vol. 93, American Institute of Aeronautics & Astronautics, 1984.

Poley, W. A., Stevens, G. H., Stevenson, S. M., Lekan, J., Arth, C. H., Hollansworth, J. E. and Miller, E. F., *An Assessment of the Status and Trends in Satellite Communications 1986–2000*, Technical Memorandum 88867, NASA.

Powell, J. T., *International Broadcasting by Satellite*, Quorum Books, Westport, CT, 1985.

Queeney, K. M., Direct Broadcast Satellites and the United Nations, Sijthoff & Noordhoff, Netherlands, 1978.

Ricci, F. J. and Schutzer, D., *U.S. Military Communications*, Computer Science Press, Rockville, MD, 1986.

Sison, M. K., Omura, J. K., Scholtz, R. A. and Levitt, B. K., *Spread Spectrum Communications*, Computer Science Press, Rockville, MD, 1985.

Slater, J. N. and Trinogga, L. A., *Satellite Broadcasting Systems: Planning and Design*, Halsted Press, Chichester, U.K., 1986.

Snow, M. S., *Telecommunications Economics & International Regulations Policy*, Greenwood Press, Westport, CT, 1986.

Snow, M. S., *Marketplace for Telecommunications: Regulation and Deregulation*, Longman, New York, 1986.

Telecommunications Dictionary and Fact Book, Center for Communications Management Inc., Ramsey, NJ, 1984.

Tunstall, J., *Communication Deregulation*, Blackwell, Oxford, U.K., 1986.

U.S. DEPARTMENT OF COMMERCE PUBLICATIONS

Telecommunications Policies in Seventeen Countries:—Prospects for Future Competitive Access, Project Managers: J. E. Cole and R. J. O'Rorke Jr., NTIA-CR 83-24, 1983.

1984 World's Submarine Telephone Cable Systems, Project Manager: R. J. O'Rorke Jr., NTIA-CR-84-31, 1984.

Telecommunications Policies in Ten Countries:—Prospects for Future Competitive Success, Project Managers: J. E. Cole and R. J. O'Rorke Jr., NTIA-CR 85-33, 1985.

U.S. International Information Services, Project Managers: J. E. Cole and R. J. O'Rorke Jr., NTIA-CR 85-32, 1985.

Present and Projected Business Utilization of International Telecommunications; 1985, Project Managers: R. J. O'Rorke Jr. and J. E. Cole, NTIA-CR 85-35, 1985.

Estimation of the Population of Large International Private Users, R. J. O'Rorke Jr., OIA-85-4, 1985.

International Teleports and their Significance for the U.S., R. J. O'Rorke Jr., 1987.

Analysis of 1987 Circuit Growth by Service Catagory, R. J. O'Rorke Jr., 1988.

Vignault, W. L., *World-wide Telecommunications Guide for the Business Manager*, Wiley-Interscience, New York, 1987.

Wu, W. W., *Elements of Digital Satellite Communication*, Computer Science Press, Rockville, MD, 1985.

UNAUTHORED PUBLICATIONS

Direct Broadcasting by Satellite, Report of a Home Office Study, H.M.S.O., 49 Holborn, London WCIV HB, U.K., 1981.

Handbook on Satellite Communications, International Telecommunications Union, Geneva, Switzerland, 1985.

International Satellite Directory, Design Publishers, 369 Redwood Avenue, Corte Madera, CA 94925, U.S.A., 1987.

World Satellite Almanac, Howard W. Sams & Co., 4300 West 62nd Street, Indianapolis, IN 46268, U.S.A., 1987.

Satellite Communications—1988–89 Satellite Industry Directory and Buyers Guide, Cardiff Publishing Company, Englewood, CO, 1988 (published every two years).

PERIODICALS

Daily
Aerospace Daily, Ziff Davis Pub. Co, New York.
Communications Daily, TV Digest, Washington, DC.

Weekly
Aviation Week and Space Technology, McGraw-Hill, New York.
Broadcasting, Broadcasting Publishing, Co., Washington, DC.
Communications Week, Manhasset, New York.
International Communications Week, Telecommunications Publishers, Arlington, VA.

Satellite Week, Television Digest, Washington, DC.

Satellite News, Phillips Publishing, Bethesda, MD.

Telecom Highlights, Paramus, NJ.

Telecommunications Reports, Business Research Publishing, Washington, DC.

Monthly

Aerospace America, AIAA, New York.

British Telecommunications Engineering, Journal of the Institute of British Telecommunications Engineers, London, U.K.

Cable and Satellite Europe, London, U.K. Cable and Satellite Magazine Ltd. A 21st. Century Publishing Company, 531–533 Kings Road, London SW10 QTZ, U.K.

Communications International, Thomson Publishing, London, U.K.

Communications News, Edgell Communications, Cleveland, Ohio.

Communications Systems Worldwide, Morgan-Grampian plc, London, U.K.

E.B.U. Review, E.B.U., Brussels, Belgium.

E.S.A. Bulletin, E.S.A., The Hague, The Netherlands.

IEE Review, IEE, London, U.K.

IEEE Communications Magazine, New York.

IEEE Journal on Selected Areas in Communications, New York.

IEEE Transactions on Communications, New York.

ITU Telecommunications Journal, ITU, Geneva, Switzerland.

Inter Media, International Institute of Communications, Echo Press, London, U.K.

Interspace: The European Satellite and Space News, 19 Regent Court, Fleet, Hants., U.K.

Satellite Communications, Cardiff Publishing, Englewood, CO.

Telecommunications, Horizon House, Transnational Data Report, TDR Services Inc., Washington, DC.

Telecommunications, Phillips Publishing, Potomac, MD.

Quarterly

Commercial Space, McGraw-Hill, New York.

International Journal of Satellite Communications, John Wiley, New York.

Space Communication and Broadcasting, An International Journal, North-Holland, Amsterdam, The Netherlands.

Telecommunications Policy, Butterworth Scientific, Guildford, U.K.

GLOSSARY

ABC. American Broadcasting Company

ABU. Asia-Pacific Broadcasting Union

ADE. above deck equipment

AFRTS. American Armed Forces Radio and Television Service

AID. Agency for International Development

AMS. African-Mediterranean System

AMSS. aeronautical mobile satellite services

ANT. ANT Nachrichtentechnik GmbH

AP. Associated Press

ARABSAT. Arab Communications Satellite Organization

ARINC. Aeronautical Radio Inc.

ARQ. Automatic R.Q. (Response to Query)

ASETA. Association of National Telecommunications Organizations for an Andean Satellite

ASIASAT. A Satellite Trade Name

ASTRA. application of space technologies relating to aviation

AT&T. American Telephone and Telegraph

ATU. Arab Telecommunications Union

AVD. alternate voice/data

AVSAT. Aviation Satellite Corporation

BAPTA. bearing and power transfer assembly

BBC. British Broadcasting Corporation

BDE. below deck equipment

BER. bit-error-rate

BPSK. Binary Phase Shift Keying

BSB. British Satellite Broadcasting

BTI. British Telecommunications International

C&W. Cable and Wireless

CANTAT-2. Canada-TransAtlantic Cable #2

CAT. Communications Authority of Thailand

CBS. Columbia Broadcasting System

CCC. Columbia Communications Corporation

CCIR. International Radio Consultative Committee

CCITT. International Telegraph and Telephone Consultative Committee

CDMA. code division multiple access

CELESTAR. Satellite Trade Name

CEPT. European Conference of Posts & Telecommunications

CES. coast earth station

CITAC. China's International Trust and Investment Corporation

CNN. Cable News Network

COMSAT. Communications Satellite Corporation

CPRM. Compania Portuguesa Radio Marconi

CSDC. called switch digital capability

DBS. direct broadcast satellite

DDS. Digital Dataphone Service

EBU. European Broadcasting Union

ECS. European Community Satellite

EGC. enhanced group call

EIRP. effective isotropic radiated power

ENTEL. Empress Nacional de Telecommunicaciones

ESA. European Space Agency

ESRO. European Space Research Organisation

EUTELSAT. European Telecommunications Satellite Organization

FACC. Ford Aerospace and Communications Corporation

FCC. Federal Communications Commission

FDM/FM. Frequency Division Multiplex/Frequency Modulation

FDMA. Frequency Division Multiple Access

FEC. forward error correction

FGMDSS. future global maritime distress and safety system

FINANSAT. Financial Satellite Corporation

FM. Frequency Modulation

FSS. fixed satellite service

GaAs FET. gallium arsenide field effect transistors

GEC. General Electric Company

GTE. General Telephone & Electronics

HDTV. high definition television

HPA. high power amplifier

IADP. INTELSAT Assistance and Development Program

IBA. Independent Broadcasting Authority

IBM. International Business Machines

IBS. INTELSAT Business Services

ICAO. International Civil Aviation Organization

ICSC. Interim Communications Satellite Committee

IFRB. International Radio Frequency Registration Board

IMCO. Inter-Governmental Maritime Consultative Organization

INMARSAT. International Maritime Satellite Organization

INTELSAT. International Telecommunications Satellite Organization

IRI. International Relay Inc.

ISDN. integrated system digital network

ISI. International Satellite Inc.

ITT. International Telephone & Telegraph Corporation

ITU. International Telecommunications Union

JCSAT. Japan Communications Satellite Company

JSCC. Japan Space Communications Corporation

KDD. Kokusai Denshin Denwa Company

KETRI. Korea Electrotechnology and Telecommunications Research Institute

KTA. Korean Telecommunications Authority

LAN. local area network

LMSS. land mobile satellite services

LNA. low-noise receiving amplifier

MAC. multiplexed analogue component

MARISAT. Satellite Trade Name (Comsat General Corp.)

MATRA. MATRA Espace

MBB. Messerschmitt-Bolkow-Blohm

MCI. Originally stood for Microwave Communications Inc. (now purely known as MCI)

MCS. Maritime Communications Systems

MDUC. Meteorological Data Utilisation Center

MESH. Consortium of 14 European Companies. Is an acronym for 3 of them: M = Matra Espace (France); ERNO (Germany); Hawker Siddeley (UK).

MMSS. maritime mobile satellite services

MSC. Management Services Contractor

MSS. mobile satellite services

NASA. National Aeronautics and Space Administration

NASDA. National Space Development Agency
NBC. National Broadcasting Company
NHK. Nippon Hoso Kyokai
NTIA. National Telecommunications Information Agency
NTSC. National Television Standards Committee
NTT. Nippon Telephone and Telegraph Company
OTCA. Overseas Telecommunications Commission Australia
OTP. Office of Telecommunications Policy
OTS. Orbital Test Satellite
PACSTAR. Satellite Trade Name
PAL. phase alteration by line
PAM. payload assist module
PATU. PanAfrican Telecommunications Union
PBS. Public Broadcast System
PSN. Private Satellite Network
QPSK. Quadriple Phase Shift Keying
RASCOM. Regional African Satellite Communications System
RCA. Radio Corporation of America
RF. Radio Frequency
RFP. request for proposal/purchase
SABC. South African Broadcasting Corporation
SAP&TC. South African Post and Telecommunications Corporation
SBTS. Sistems Brasiliero de Telecommunicacoes por Satellite
SCPC/PCM/PSK. single channel per carrier
SDUC. Secondary Data Utilisation Centres
SECAM. sequence a memoire
SES. ship earth station
SES. Societe Europeene des Satellites
SIN. Spanish International Network Inc.
SITA. International Society for Aeronautical Telecommunications
SMS. Satellite Multi Services
SPADE. single channel per carrier PCM multiple access demand assignment equipment
SPAR. SPAR Aerospace Ltd.
SPO. Satellite Planning Office
SS-TDMA. satellite switched time division multiple access
SSOG. Satellite System Operational Guide
SSPA. solid state power amplifier
STC. Sudan Telecommunications Corporation

TAT. Trans-Atlantic Telephone cable

TDMA. time division multiple access

TRD. technical requirements document

TRW. Originally stood for Thompson, Raymo & Wallridge. Now known only as the Company TRW Inc.

TTC&M. tracking, telemetry, control and monitor

TVRO. Television Receive Only

TWTA. traveling wave tube amplifier

UNDP. United Nations Development

USIA. United States Information Agency

VISTA. Name of an INTELSAT service

VOA. Voice of America

WARC. World Administrative Radio Conferences

WUTCO. Western Union Telegraph Company

INDEX